The ABZs of Birds

Your Everything Guide to the Mystery and Magic of Birds

by **Kathryn Hornbein**

illustrated by **Cynthia Hunt**

old crow books

The ABZs of Birds: Your Everything Guide to the Mystery and Magic of Birds
Copyright © 2025 by Kathryn Mikesell Hornbein

Printed in the United States of America

Illustrations by Cynthia Hunt

Old Crow Books
1317 Devils Gulch Road
Estes Park, CO 80517
www.theabzsofbirds.com

LCCN: 2025912778
ISBN: 979-8-218-70918-1

www.theabzsofbirds.com

Dedication from Illustrator
To Irene, who taught me how to fly.

Dedication from Author
To Hazel Wren and in memory of her grandpa Tom, my husband.

Acknowledgments

Infinite thanks to Cynthia Hunt, friend and illustrator-plus. Her heart and soul resided in the Indian Himalaya for 35 years where she did village-development work, including writing and illustrating books for village kids. When she was stuck in the United States one winter, I leaped in and asked if she'd like to illustrate my bird book. I don't think she knew a Steller's jay from a flicker, but she learned about North American birds and taught herself how to draw them. And she had ideas. She suggested I create "subject" pages like the ones on flight and feathers. Crow emerged whole, one day, from her witty, creative brain. Supposedly dyslexic, she argued the nitty-gritty niceties of grammar and writing style—and was usually right. Now and again, whimsical paragraphs mysteriously appeared in my text. (Dandelions and dog-calling contests? Whaat??) And, crucially, she hung in with our book when her physical presence and priorities were in faraway Ladakh.

Though the Cornell Lab of Ornithology doesn't know it, I owe them a great deal. When I retired from pediatrics and was feeling an intellectual void, I happened on their rigorous *Home Study Course in Bird Biology.* Though all my errors are my own, I returned to their resources—the coursework and website *Birds of North America* (now *Birds of the World*) again and again while writing the text. Other frequently used resources: Field guides by National Geographic (Dunn and Alderfer), the National Audubon Society (Sibley), *National Geographic Kids Bird Guide of North America* (Alderfer), and *Smithsonian Kids' Field Guides: Birds of North America East.* Also: *The Sibley Guide to Bird Life & Behavior; The Birder's Handbook* by Ehrlich, Dobkin, and Wheye; *Nests, Eggs, and Nestlings of North American Birds* by Baicich and Harrison; DK Eyewitness Books' *Bird* by Burnie; *The Audubon Society Encyclopedia of North American Birds* by Terres; *Birder's Dictionary* by Cox; and numerous online resources.

Dan and Barbara Gleason reviewed the book for accuracy. Freelance editor Chris Frisella and Luminaire graphic designer Claire Flint Last (and Cynthia, of course), reviewed the text, artwork and layout. They sent me back to my resources time and again. I learned a lot from them. Many thanks!

Many people have leafed through the manuscript, captivated by the illustrations. Charlie Roach, now all grown up, could not forgive me for using pelican instead of penguin on the "P" page. He wasn't at all mollified when I pointed out that this was a book of North American birds, and penguins hadn't migrated up here yet. Sorry Charlie.

My husband Tom stood by me, believing in me through thick and thin for fifty one years. Thank you T, I miss you.

TABLE OF CONTENTS

That weasel Crow

Here's how Crow weaseled his way into this book.

I'm using the alphabet to introduce you to fascinating things about birds. You know, A is for…

"CAW! CAW! Where am I in your book?"

What?! A crow?
I nearly fell out of my chair.

He peered down at me from the top of my computer screen!

"Cardinal is my 'C' bird, not crow," I said, trying to recover. "And for good reason. Everyone loves cardinals. Crows come in noisy, obnoxious flocks. People don't like them. Think scarecrows."

"We eat a little corn, and lots of nasty insects. Farmers should love us," said Crow, preening his glossy feathers. "And besides, cardinals only live in the East. We're everywhere."

"So? You're too much of everywhere. You bully smaller birds and eat their eggs and chicks."

"Those bitsy things are a tiny part of our diet. Besides, we have to feed our babies!"

Whoa! "You kill babies to feed babies? That's disgusting!"

Crow balanced on one foot, scratching his head with the other. "Who takes better care of their kids? Cardinals kick theirs out of the nest to have a second family. We keep ours around. We watch over them, teach them. Then they help with next year's kids, like a real family."

"Crows are always
making mischief,"
I said.

No way was I ejecting
cardinals from this book!

"What's the harm of a little
mischief? A pulled tail here
and there…" said Crow.
He pooped, just missing my tea.

"Hey!" I yelled.

"Crows play," he said, cocking his head and fixing
his beady eye on me.

"You're changing the subj—"

Crow gave a little hop, his claws scratching my
computer screen.

"Have you seen cardinals sliding down a
snowy slope or playing tug-of-war with a stick?
Do cardinals catch a breeze and turn cartwheels
across the sky? Do cardinals imitate your
stupid cell phones?"

"And? Your point is?" I asked. My phone rang.
I reached for it; no one was there.
Can crows smile?

Cynthia, the illustrator, looked over my shoulder
at my screen. "Of course Crow has to be in our
book," she said. "In fact…well…he mysteriously
shows up on every page I draw…"

"Holy Guacamole!" I surrendered. That's how
Crow ended up everywhere in this book.

Smart. Mischievous. Persistent. That's Crow.

3

Wind. The Albatross can't fly without it.

He soars endlessly on ocean winds, wings hardly moving. Without wind, he can't flap his long, narrow wings powerfully enough to keep his heavy body in the air.

In the water below him he sees squid and splashes down, snagging them with his long hooked **beak.*** After eating, he flaps his wings vigorously and races on webbed feet over the ocean's surface into the wind (a headwind) until he's finally aloft.

If you walk against the wind you struggle. Not a bird taking flight! As birds launch themselves against the wind, air flows around their bodies, lifts them, and they become airborne.

The albatross has been gone several days, and it's time to change shifts. He flies toward his nest four hundred miles away, expertly using headwinds, tailwinds, crosswinds, and the updrafts from the ocean waves to hurry him along.

Gee! How graceful!

OOMMPH!!

Eight hours later he arrives at a tiny Hawaiian island and coasts past dozens of identical nests in the **colony** to his own. He could use a good headwind to brake himself. The wind is from his back, though, and (whoops!) his flight ends in a somersault.

Unfazed, he hops up and greets his **mate** and chick. When the fuzzy chick nibbles his **bill**, he vomits (regurgitates) squid plus nutritious stomach oil into its throat.

Yum, breakfast.

Now the mother's off to sea to forage (hunt food).

Besides their diet of squid and fish eggs, albatrosses sometimes follow boats and **scavenge** their discarded garbage, which they feed their chicks. Just before they leave the nest, chicks cough up things they can't digest, like squid beaks and fish bones. Sometimes it's plastic bags and other boat rubbish. That trash can be deadly.

At five months old, the chick looks like its parents, but is heavier. Its parents have fed it well to give it a good start in life. It leaves the nest, and with the colony's other chicks goes to the water and learns to swim, fly, and find food.

If a tiger shark, one of its natural enemies, doesn't grab it, the young albatross is off to sea for several years. A **pelagic** life is normal for albatrosses, who fly thousands of miles in search of food.

*Definitions for **bolded** words appear in the glossary beginning on page 73.*

You are standing ankle deep in a squishy marsh.

Something is hiding in the tall grasses.
Can you see it? It looks like marsh grass
swaying in the wind. But is that a brown
neck and a beak pointed toward the sky?
Is that a bird, hidden among the
grasses? You creep forward
to see better.

"KOK! KOK!"

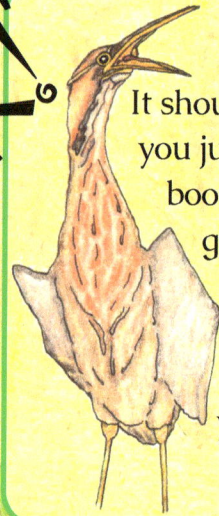

It shouts so loudly
you jump out of your
boots. Not marsh
grass at all,
but a striped
brown bird
with a long
yellow bill.

A *B*ittern!

The bittern gulps in air and makes an even weirder sound, loud and low and booming, like

PUMP - ER - LUNK!

It's so loud it startles people a half-mile away. But it's just a bittern making itself heard through all the reeds and grasses.

And its eyes? Eerily, they're fixed right on you.

A bird's eyeballs don't move in their sockets like ours do, so it has to move its head instead. (Watch the birds in your yard.)

A bittern's eyes are placed so that when standing with its beak pointed skyward pretending to be marsh grass, its eyes are on you (or a **predator**), watching what you're up to. When hunting, its beak points straight ahead, and its eyes point down, searching for its next meal.

To forage, the bittern stalks stealthily through the high grasses of the marsh looking for

a frog,

dragonfly,

fish,

or anything

that moves.

When it sees some little animal, the bittern slowly snakes its head back and forth to see it better, then darts its beak forward and grabs it, gulping it down head first, bones and all. It coughs up the bones or shells in a neat package called a **pellet**.

Are you really lucky?

If so, out of your window you'll see a bright red bird with a **crested** head and a black mask, and hear a lovely song.

It's a male northern

Cardinal.

People used to think that it was mostly male birds that sing, but we now know that many females sing too. Male and female cardinals sing to each other.

The female cardinal, who is beige and gray with red trim, hides her nest (called a cup nest) among the leaves in trees or vines. She takes twigs her mate brings her, breaking them and bending them around her body, turning round and round to shape the nest like a cereal bowl.

She lines it with layers of soft leaves and grasses. And then, she lays three speckled eggs. Her mate feeds her as she starts their family.

Once the babies hatch, the fun begins. The nestlings are almost featherless, and completely helpless. Their eyes are stuck shut. They lift their heads, gape their mouths open wide, and peep to be fed.

The parents fly miles and make dozens of trips to the nest, bringing insects and stuffing them down wide-open beaks. The female broods the babies (covering them with her fluffy body to keep them warm), and the parents even fly poops away from the nest.

The nestlings grow so fast that in only ten days their eyes are open, they are covered with feathers, and they even have little crests on their heads. One morning they perch on the edge of their nest. Their parents call to them and offer them juicy insects—from a distance. The chicks flutter to the nearest branch and teeter there, begging to be fed. Little by little they learn to fly better and find their own food.

When they're old enough, guess what? Their parents start another nest. More babies!

What ARE you doing?

A cardinal rap, man! It's a cardinal rap!

Different birds—Different nests!

Mud pellets

Cliff ledge

Burrow

Eggs and Babies

Eggs! Different sizes, shapes, and colors, depending on what bird is growing inside and the type of nest. Some eggs are colored in a way that hides (**camouflages**) them in plain sight. Eggs in nest cavities are often white. They're out of sight, not camouflaged.

- 14

- 21

- 28

The reason for these differences is to give the egg— and baby—the best chance of surviving.

The female lays an egg each day or two. Some birds lay one egg (albatrosses), some three or four (cardinals), some a dozen (ducks). Depending on the kind of bird, some eggs hatch in ten days, and some take weeks.

Birds sit on their eggs (incubate them) to keep them at exactly the right temperature. They lose the feathers in a patch on their stomach (an incubation patch) so the eggs are nestled against their bare skin for warmth.

Some birds, like ducks and cardinals, wait to incubate their eggs until they've all been laid. Their babies all hatch together, and are the same age..

Birds like owls and eagles begin incubating from the time the first egg is laid, and their babies hatch on different days. If food is scanty, the younger chicks die, which saves the stronger, bigger chicks. It seems cruel, but the job of the parents is to successfully raise at least part of their family.

New baby birds are called hatchlings. Hatchlings have an "egg tooth" on their beak—their only tooth, ever, and it's not even a real tooth. They use it to peck their way out of their shells, then it falls off.

Hey! Do you all all have the same birthday?

- 7

- 1

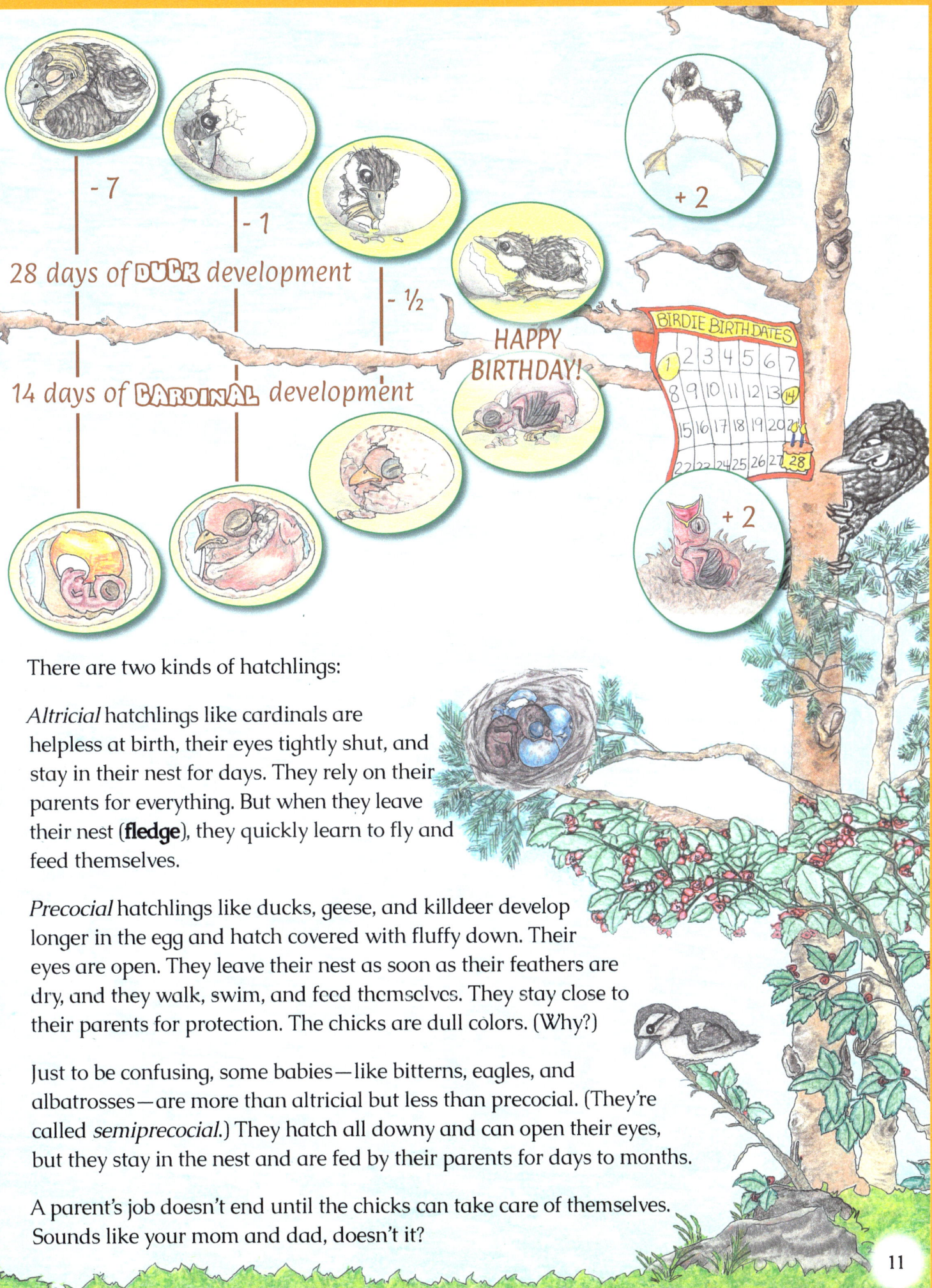

28 days of DUCK development

- ½

14 days of CARDINAL development

HAPPY BIRTHDAY!

+ 2

BIRDIE BIRTH DATES

1	2	3	4	5	6	7
8	9	10	11	12	13	14
15	16	17	18	19	20	21
22	23	24	25	26	27	28

+ 2

There are two kinds of hatchlings:

Altricial hatchlings like cardinals are helpless at birth, their eyes tightly shut, and stay in their nest for days. They rely on their parents for everything. But when they leave their nest (**fledge**), they quickly learn to fly and feed themselves.

Precocial hatchlings like ducks, geese, and killdeer develop longer in the egg and hatch covered with fluffy down. Their eyes are open. They leave their nest as soon as their feathers are dry, and they walk, swim, and feed themselves. They stay close to their parents for protection. The chicks are dull colors. (Why?)

Just to be confusing, some babies—like bitterns, eagles, and albatrosses—are more than altricial but less than precocial. (They're called *semiprecocial*.) They hatch all downy and can open their eyes, but they stay in the nest and are fed by their parents for days to months.

A parent's job doesn't end until the chicks can take care of themselves. Sounds like your mom and dad, doesn't it?

Ducks have webbed feet

for swimming, like albatrosses.

A dabbling duck like a *mallard* eats off the water's surface or from the shallow bottom of a lake or stream. It tips upside down, its tail pointing to the sky, and stretches to grab underwater plants and **invertebrates**.

The male mallard has a splendid green head. The female is mostly brown. Like the female cardinal, her dull color camouflages her, so she isn't easy to see as she sits on her nest; many predators would like to make a meal out of her eggs or babies—or her.

What is a diving duck? It's a duck like a *bufflehead* that dives and swims under water, hunting pondweeds, insect **larvae**, **seeds**, **crustaceans**, and small fish.

Bufflehead winter on lakes or protected ocean bays. An adult bufflehead can drink salt water. But for a young bufflehead (or you), it is undrinkable, so bufflehead parents nest along freshwater lakes or ponds.

A mallard nests on the ground. Not the bufflehead. That mom lays her plain white eggs inside an old flicker hole in a tree *(see p. 60)*. They use nest boxes too. When the babies are only two days old, they hop out of their nest, flutter down to water, and paddle away after their mother.

Ducklings always feed themselves, but their mother finds their food with them at first.

Bufflehead ducklings dabble like mallards for a few days until they learn to dive. Mothers brood their babies at night for a couple weeks, especially in cold weather. Female mallards remain with their ducklings until they learn to fly. Buffleheads are on their own before they can fly, and collect in groups together. Male ducks don't hang around to help with parenting.

In the late summer, adult ducks can't fly for several weeks while they molt (shed) their worn-out flight feathers and grow new ones. Flightless mallards all look like females, brown and camouflaged from their hungry enemies. You might think all the males have disappeared. Then they molt again, the males grow their bright new feathers, and off they fly.

There are many kinds of ducks. But if you see something that doesn't look quite like a duck, maybe it's a goose or a swan, or a loon! These fathers do stay around to help out with their young.

Color's great. But how do I grip a perch?

13

Is a bald Eagle bald?

You'd expect a bald eagle to be bald, but it has bright white head (and tail) feathers. And a golden eagle should blaze in the sunlight, right? But it's mostly brown with some yellow feathers on its neck. Birds' names don't always match their appearances.

Speaking of expectations, you'd expect more from mighty eagles than their squeaky cries.

Their cries may be wimpy, but eagles' **courting** is amazing. Courting bald eagles fly high into the sky, lock their **talons (claws)** together, and tumble through the air until you think they're going to crash. At the last minute they separate and fly high again. Golden eagles court with swooping dives and chases.

Bald eagles hunt from tall trees near water because their favorite food is fish. From their high perch, they spot a fish with their incredible eyesight, swoop down, and grab it in their strong talons. They are **opportunists**, though, and if they can pirate food from another animal they will. They eat carrion (dead animals), waterbirds, and small animals—anything they can nab.

Is a golden eagle golden?

Golden eagles live in open areas like prairies, with nearby cliffs for nesting. They chase after their favorite foods, hares and other smallish mammals, twisting and turning to follow them as they try to escape. Sometimes they land on a larger animal like an antelope and ride it until it dies of exhaustion.

Eagles' talons and big hooked beaks are their knives and forks to tear their meal into pieces before bolting it down. They cough up bones, feathers, and fish scales in pellets.

Eagles' nests (aeries) are out of reach of most predators—think of the golden eagles' cliff nests. Bald eagles make enormous nests in treetops, adding to them over years. They can weigh a ton! Can you imagine a one-ton car sitting in a treetop for years?

A young bald eagle grows as big as its parents in a few months, but it won't look like them for three or four years. Until then it's blotchy brown, more like a golden eagle.

Eaglets strengthen themselves before they fledge by flapping their wings, holding tight to their nests so they won't take off before they're ready.

Would eaglets ever leave their nests if they knew what a L-O-N-G way it was to the ground?

You've got a built-in knife and fork. I don't.

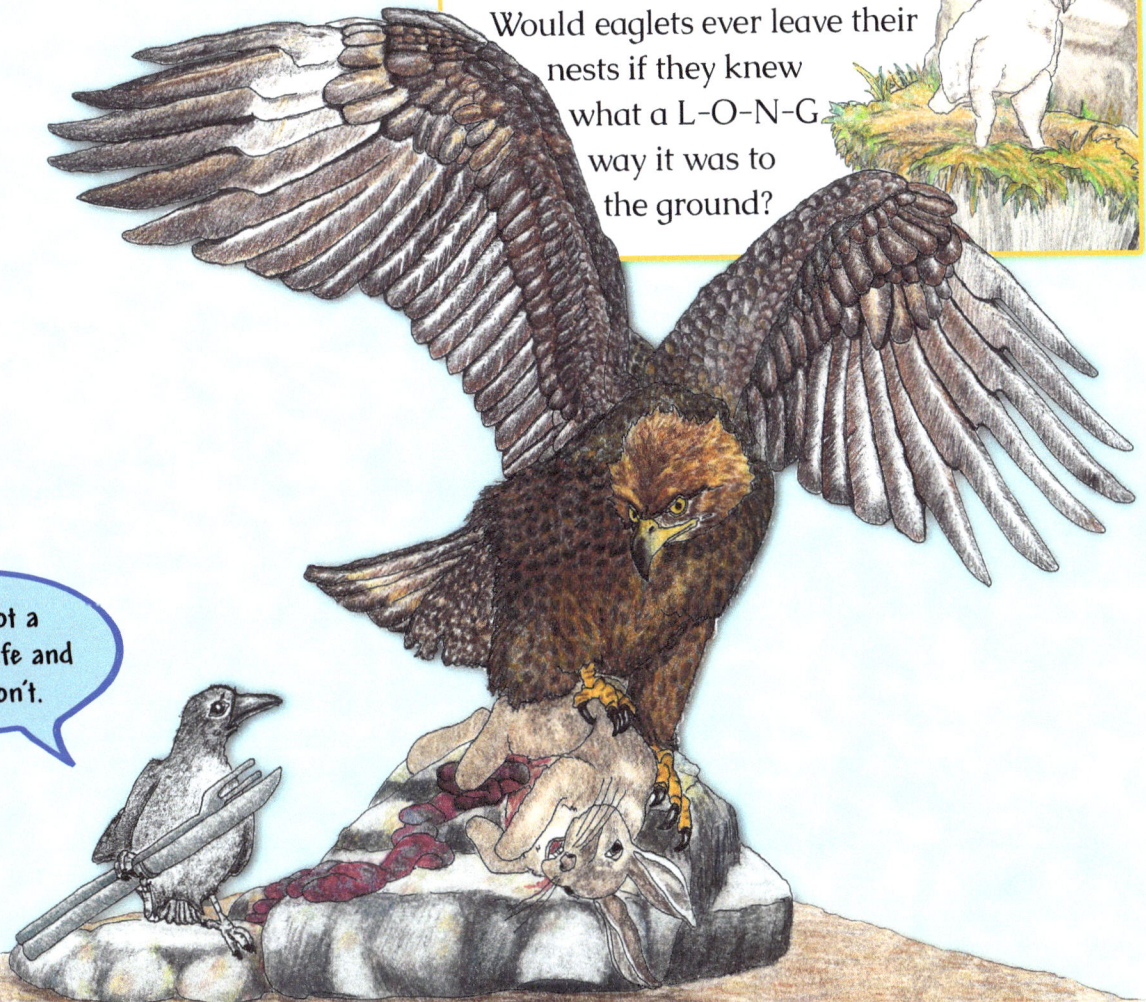

15

We have noses, mouths, hands, and feet. Birds have
Beaks and Claws

Beaks? Bills? They mean the same thing. And birds breathe through their nostrils, which open into their beaks.

Beaks tell us about what birds eat:

 Short, strong, triangular beaks for seed eaters that pluck and shell seeds.

 Short, slender beaks for insect eaters who pick bugs from leaves and branches.

 Broad bills, or small bills with large mouths, and often whiskers, for insect eaters who snatch **prey** from the air.

 Powerful beaks for woodpeckers pounding and excavating holes in trees.

 Tough, hooked beaks for meat eaters like **raptors** that tear their prey to pieces.

 Tiny jagged **serrations** on bills of **waterfowl** that grasp plants or wiggly prey

 Slender bills that sometimes have **sensors** at the end for **waders** or **shorebirds** probing mud for food.

 Long, slender bills that slide deep into flowers for hummingbirds drinking **nectar**.

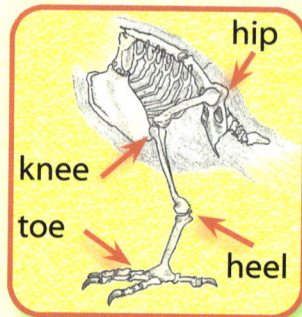 Strong bills for **omnivores**, such as crows and gulls, who eat many kinds of foods.

Birds walk on their toes. What look like backwards knees are actually their heels. Their knees are hidden in their feathers and bend forward like ours do.

Feet tell us where birds live and how they get food:

 Swimmers like ducks and gulls have webbed feet.

 Waders have long spidery toes to walk on top of squishy mud or even lily pads.

 Raptors have strong feet and talons for catching, killing, and carrying their prey.

 **Galliforms**, like quail and turkeys, have rugged feet for running and turning over leaves and soil as they search for food..

 Hummingbirds, swallows, and swifts have small feet. They spend most of their time in the air and no time hopping.

 Perching birds like chickadees and robins have three forward toes and one pointing backwards (like our thumb), for landing on a branch and holding on, even while they sleep.

 Woodpeckers don't perch. They usually have two toes forward and two backward, for clinging and climbing. Their stiff tail feathers help prop them on tree trunks.

What else can birds do with their beaks and feet?

Dig, carry, build nests, care for their babies, groom their feathers, scratch, fight, go courting, sing. And, remember the egg tooth? Pretty good, for no hands.

Who did it?

A drumroll from a tree.
A half-eaten hot dog.
A frog leg, but no frog!
A scratched-up garden.
Duckweed and fish
gone from the pond.
No mosquitoes.
Sunflowers minus
their seeds.
Nectar lapped from
the honeysuckle.

Who are the thieves in these crimes?

Their beaks and feet will tell you! Look at the birds in the lineup and match them with their crimes. (See p. 72 for answers.)

1

2

3

4

5

6

7

8

9

10

11

17

Flamingos

used to live by the thousands in Florida, but they were killed off in the 1800s for their eggs and gorgeous feathers. Now they are—maybe—returning, flying north from the Caribbean Islands or Mexico.

Flamingos are a magical color, *pink*, from their diet of tiny shrimp and algae. These foods contain carotenoids, which are red or orange chemicals. Without carotenoids, flamingos aren't pink—or healthy.

A flamingo is a most unusual bird. It has long skinny legs. Its neck is long too, and it eats down by its feet.

To eat, it holds its breath, and sticks its head

UNDER water,

UPSIDE down,

and **BACK** -wards.

It sweeps water and mud into its bill, keeps the food, and strains everything else back out. How does its oddly shaped, fringed bill and amazing tongue manage to separate tiny foods from mud?

Flamingos live in colonies (called flamboyances, a **collective noun**) around freshwater or saltwater lagoons and mudflats. They fly together like pink clouds. They often stand on one leg—to conserve heat or use less energy while resting? No one is sure why.

When it's time to **breed**, flamingos honk, swivel their heads, and march around all together. This strange dance seems to **synchronize** their nesting, so all the chicks hatch at the same time. But that means a disaster like a severe storm can destroy the year's **offspring**.

Parents build a nest that resembles a sand mountain, with a dent in the top for their one egg, which hatches in a month. When the chicks are a few days old, they go to day care, called a crèche. Penguins, pelicans, Canada geese, and other birds use crèches too. Some of the adults watch the babies while the rest eat.

Newly hatched chicks have straight bills and are white or gray; they haven't eaten their carotenoids yet. Both parents make bright red crop milk deep in their throats, and drip it from their bills into their chick's mouth for weeks, until it can feed itself. (Two other birds make crop milk—pigeons and penguins.) After several years, the chicks have grown up—and turned pink.

...and no quick trips for milk!

People call them "seagulls" because they think only live around the sea (ocean). Gulls

Maybe, though, they should be called "water gulls" because they live on freshwater, too—lakes, rivers, and wetlands (squishy, wet and swampy land).

Or "garbage gulls," because you'll find them scavenging wherever garbage collects, like fast-food parking lots, city dumps, and picnic areas. They have **adapted** to humans very well.

Like other seabirds, they can drink salt water. All the extra salt collects in a salt gland above their bills, and very salty water dribbles out their nostrils, like a cold that won't go away.

Gulls are omnivores like crows. Look at what they eat (alive or dead)

- Fish, shellfish
- Birds, birds' eggs, baby birds
- Rodents and other small mammals (mice, prairie dogs, rabbits)
- Amphibians (salamanders, frogs, toads)
- Reptiles (turtle eggs and babies, lizards)

- Earthworms, **grubs**, insects, slugs
- Seeds, berries
- Animal poop
- Carrion
- Garbage, of course, and any people food they can beg or steal.

It's Mine!

No! I'm havin' this my way!

Gulls swallow their food whole, and regurgitate pellets. Their pellets contain bones, feathers, crab claws, paper, even plastic and glass! Gulls aren't always smart about what they eat.

But they are clever. Like crows, they drop shellfish on a hard surface to crack them open. (Young gulls drop golf balls, thinking they'll find goodies inside.) Gulls chase other birds, even eagles, and steal their food.

There are many **species** (types) of gulls. The brown gulls you see aren't a separate species though. They're kids. Like eaglets, they're soon as big as their parents and on their own, but they don't get their adult colors of white, black, and gray for several years.

Gulls' feet are webbed; they swim and make shallow dives. Watch them fly— they're so graceful as they soar and swoop, catch a breeze and float on it, land on the water, and take right off, without having to run on the surface like many waterbirds.

Listen to their mewing calls. They remind you of the ocean, even if they're stealing food from you in Kansas.

21

Flight

How come birds can fly and you can't?

Gravity keeps you (and birds) from flying off into space. Jump. You're back on the ground in a second.

Then there's friction, called drag, which makes it difficult to move something against something else, like dragging a heavy box over grass. There's even friction between you (and birds) and air.

How then do birds fly?

Birds have ways of overcoming gravity and drag, and they add some **evolutionary** design-genius extras. They have:

lift

thrust

drag

weight

Thrust, which is power. You use your legs to run. Cars and airplanes use motors. Birds flap their wings.

Design. Birds are **streamlined** to decrease drag. They're torpedo-shaped. Their feathers "zip" together to make a smooth surface, thanks to barbules. *(See the pages on feathers to find out about barbules.)*

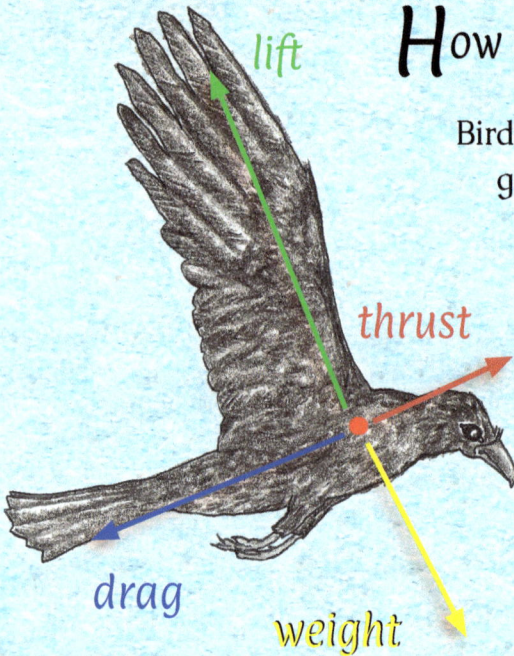

Lift. Birds' wings (and airplane wings too) are a special shape, called an airfoil, which is more curved on the top than the bottom. Because of this, air rushes over the top of the wing faster than under it, and lifts it. (Kind of like magic, isn't it?)

If they have a choice, birds and airplanes take off against the wind, which creates even more lift because of the increased rush of air over their wings.

AIRFOIL

Light weight. A hummingbird weighs less than a nickel. A bald eagle, seven feet from wingtip to wingtip, weighs only ten pounds, a little more than a human newborn.

Instead of heavy noses, jaws, and teeth, birds have lightweight *beaks.*

Birds have fewer *bones* than mammals their size.

Many bird bones have bubble-like air spaces which lighten them—an adaptation more common to soaring birds than deep-diving birds. (Why?) Their feather shafts are hollow, too.

Like us, birds have two *lungs* for breathing. They also have nine amazing *air sacs,* like tiny balloons, tucked inside their bodies that connect to their lungs, and boost their breathing and buoyancy.

Birds lay *eggs* instead of carrying growing babies inside their bodies.

flight muscles

Wings. Birds have wings instead of arms. Huge muscles, from a bird's wings to its breastbone's *keel,* power strong flight. Fifty muscles in each wing control flight. Tails help steer and brake.

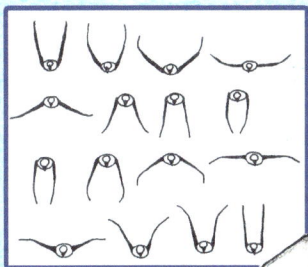

Thrust Cycle

Watch birds' wings. Can you see their thrust cycle?

It's the cargo load that gets me!

So why can't we fly?

With our heavy human bodies, we'd need wings so big we wouldn't be strong enough to flap them. And where would we be without arms and hands, anyway?

Zip up, down

and backwards.

Zip left, right

and hover.

Wings blur
        like propellers.
Our helicopter bird.

No other birds fly like

Hummingbirds.

Their wings flap *fifty times per second*, they can fly thirty mph, and turn on a dime.

They are the world's smallest birds. The tiniest species weighs less than a penny, is barely bigger than a bumblebee, and sometimes is attacked by dragonflies!

Why does this tiny bird hover around flowers? A hummingbird sticks its beak deep inside a flower. With its long tongue, it laps the flower's nectar for energy. At the same time, the flower's powdery **pollen** falls onto its head. Some pollen falls into the next flower the bird visits. That's called pollination, and it helps plants reproduce—that is, make more of themselves. Hummingbirds and flowers benefit each other, an example of **symbiosis**.

Hummingbirds also eat tiny bugs, insect eggs, and larvae, and they feed them to their babies. Sometimes they eat the bugs out of a spider web, and then maybe they eat the spider too.

I'm a helicopter bird too!

Though hummingbirds don't sing beautiful songs like cardinals, they communicate very well together. They chatter in squeaks. The male flashes dazzling colors from the **iridescent** feathers on its head and neck (their gorgets, pronounced *GORjets*). They whiz and dive in looping patterns when they're courting or claiming their **territory**.

A hummingbird nest fits in your hand. It's made from feathers and dandelion fluff held together with spider webs, and covered with bits of bark to camouflage it. The eggs look like white jelly beans. The mother incubates them and cares for her chicks by herself. She will dive-bomb anything, even *you*, if she thinks they are in danger.

Most North American hummingbirds **migrate**. In summer they nest here, in our gardens, deserts, mountain meadows, and forests. In the fall, they fly south where it's warm, and there are more flowers and insects. Some migrate along land, feeding as they go. The ruby-throated hummingbird migrates 500 miles over the Gulf of Mexico in under 24 hours!

The summer's babies migrate after their parents have already left. How do they know where to go? Nobody knows for sure, but it must involve **instinct**— **behavior** an animal **inherits**, and doesn't have to learn.

If you live in a city, or on a farm, or play in a park, watch for small brown birds that hop around, pecking at crumbs, and cheeping. These are *house sparrows*. That monotonous *cheep, cheep, cheep* is their only song.

Do you see **flocks** of noisy black birds with white-spotted feathers that love lawns and walk instead of hop? They're *European starlings*. In flight, their huge, swirling, shape-shifting flocks (called murmurations) are mesmerizing. And noisy!

These birds, and dozens of other species, were

Introduced

to North America from other parts of the world by people.

Some were imported and released by immigrants homesick for familiar birds. Some were introduced to eat pesky insects, and became pests themselves. Others were brought in as pets, for zoos, or to be hunted, and then escaped.

Most introduced birds don't survive, because life here is too different, and they can't adjust. (What if someone dropped you at the North Pole without any food or warm clothes?) Some couldn't find mates to start a new population.

Some birds survived and thrived; one hundred European starlings and a few house sparrows were introduced here in the late 1800s. Now there are two hundred *million* starlings and twenty five million house sparrows all across North America.

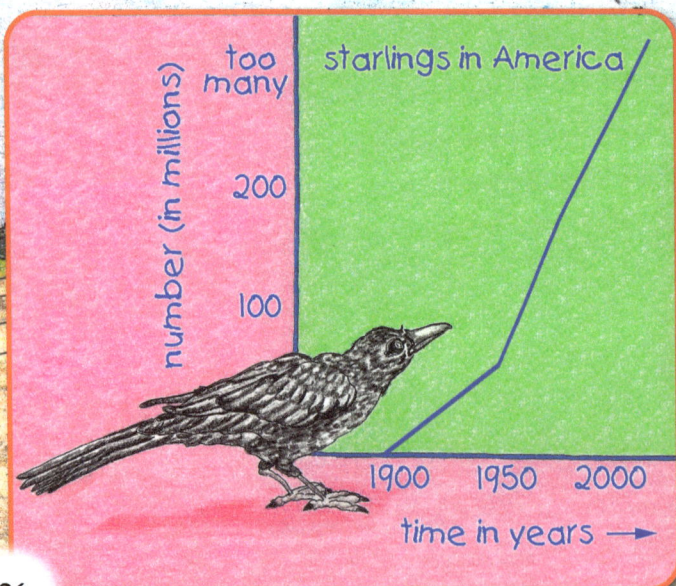

number (in millions) — starlings in America

too many
200
100

1900 1950 2000

time in years →

Introduced species have few natural enemies to keep their populations in control. They become out-of-control pests (called invasives) and take over **native** birds' territories, eat their food, and use their nest sites.

In any big city, open a window and look at the ledge. Bird **droppings**? Pigeons (officially called *rock pigeons*) love cities. New York City has over a million. Each bird produces almost fifty pounds of droppings a year—fifty million pounds a year for the city to cope with—and that's just the pigeons!

On farms, starlings eat seeds and crops. They can kill trees with their huge **communal roosts** and toxic poop!

Large flocks can even get sucked into jet engines and cause trouble for planes— and the birds.

Jays!

Jays! Blue jays, Canada jays, Steller's jays, pinyon jays, scrub jays, Mexican jays, and Clark's nutcrackers— all are different types, or species, of jays.

Jay species are alike in many ways.

Their feather colors are mixtures of blue, gray, black, or white. Two species, the blue jay and Steller's jay, have crests.

Like crows and gulls, they're omnivores. They eat insects, nuts and large seeds like acorns, fruit, eggs, small animals, carrion, and people food. Gray jays and Clark's nutcrackers are called "camp robbers." They hover around picnickers, hoping for a handout or maybe grabbing a cookie on a flyby.

Jays cache (store) food for winter, when food is scarce. They remember where they've hidden thousands of seeds. Some seeds get forgotten and grow into new trees.

Jays are social, with strong family ties. They're bold, aggressive, intelligent, and inquisitive.

They're raucous and talkative, always squawking about something. They're mimics; if you hear a hawk or cellphone, it could be a jay! At their nests they sing sweet, private songs for their families.

Jays mob owls and hawks (**birds of prey**), calling other jays in to circle around, making an enormous racket. If the poor bird gets disgusted and tries to escape, the screaming jays follow it like a noisy tail. Listen for mobbing jays; you might see a fleeing hawk!

Jays mate for life, like a marriage. But if their mate dies, they find another. Like all birds, their job is to have many offspring, so their species survives, because most baby birds die. They're killed by other birds and wild animals (who have to eat and feed their own babies) and by house cats, or they starve, or fall out of their nest too soon, or freeze in nasty weather.

Think he has a treasure map?

Must have a GPS.

Jays and crows are **corvids**; they're relatives. How are they alike? Which jay species are your neighbors? You'll have at least one!

Their name makes them sound like deer killers. But no,

Killdeer

are robin-sized birds that eat worms, bugs, snails, and not deer.

They call *kill-deer, kill-deer,* day and night, in a high voice you'll remember once you've heard it. (Funny how we like to hear "our" words coming out of birds' beaks!)

long way down...

Killdeer are brownish-gray and white, with two black necklaces and long running legs—run, stop stock-still, run, pounce. And with their narrow, pointed wings, they're equally agile in flight.

They are shorebirds, but they've adapted to human creations like man-made fields, road shoulders, and lawns all across the country. They follow tractors, eating the grubs and insects that get stirred up.

The parents make a nest by heaping gravel or stones in a low circle on the ground, or—oh, my—on gravel roofs.

The precocial chicks look like downy little adults, except they have only one black necklace. They swim before they fly, and they feed themselves from the get-go.

Killdeer parents defend their eggs and chicks from an enemy (like you, or maybe a fox) by rushing right at it, flapping their wings, and calling loudly to scare it away. Or a parent flutters away from the nest, dragging a wing as if it's hurt. When the fox decides an injured bird is easy pickings, it chases the parent, who suddenly flies away!

When you see that distraction display, you know there are eggs or chicks nearby. Being careful where you step, search for them. They are camouflaged to look like stones and gravel, and are hidden in plain sight. Can you find them?

Birdwatchers call them
"Little brown birds,"

but actually they have mixtures of quiet colors like rust and brown, white, black, and gray. Some may have spots of bright colors, and fancy trim, but you'd generally call them plain.

Small and inconspicuous (camouflaged), they flit among trees and skulk through weeds and shrubs—quite the opposite of our flashy hummingbirds.

In autumn and winter, if they don't migrate, you'll find them traveling together as mixed-species flocks (feeding flocks), twittering and peeping to one another (contact calls) as they forage for seeds and insects.

Can you find these little birds?

Most *wren* species are small and brown, with loud songs. They hide themselves well, so it helps to learn their songs. With their slender, curved beaks, they search out insects and spiders. They don't come to feeders. (Why not?) They have unusual behaviors; they make **dummy nests** and some wren species sneak into nearby nests and destroy the eggs or nestlings— to ensure that they can find enough food for their own offspring?

crowded yard...

The fifty species of North American *sparrows* are a hundred shades of brown and gray, often striped, with short seed-crushing beaks. They forage among shrubs and grasses. Some species sing beautifully, others in buzzes and cheeps. You can often tell sparrows apart more easily by their songs than their appearances.

Chickadees have dark caps and bibs, and white below their eyes and on their bellies. If they're scared, they call *chick-a-dee, dee, dee*. The more worried they are, the more *dees*. Other birds and animals hear them call and check for lurking danger. Some species have a sweet spring whistle.

Siskins look like small brown-striped sparrows, but when they open their wings you'll see yellow feathers. They are noisy and greedy, and travel in flocks looking for thistle and sunflower seeds—and feeders, where they eat nonstop. But their greed has a reason; siskins burn more energy than other birds their size.

Snowbirds. That would be the *juncos*, since they winter all over the US, and visit our feeders. Juncos have dark hoods and wings, gray or rust backs, white bellies—and pink beaks. In flight they flash their white outer tail feathers. Their spring song is a sweet trill.

Nuthatches, finches, vireos; there are many more species of these little plain birds to discover.

Wait til the robins arrive.

33

If you had this kind of night, would you be able to sleep?

Chirpery, chirpery, chirpery sings a bird in the dark outside your window. Another sings *tweeter, tweeter, tweeter;* followed by a third: *cheeriup, cheeriup, cheeriup.* The gate squeaks. Is someone coming in? You yank your pillow over your ears.

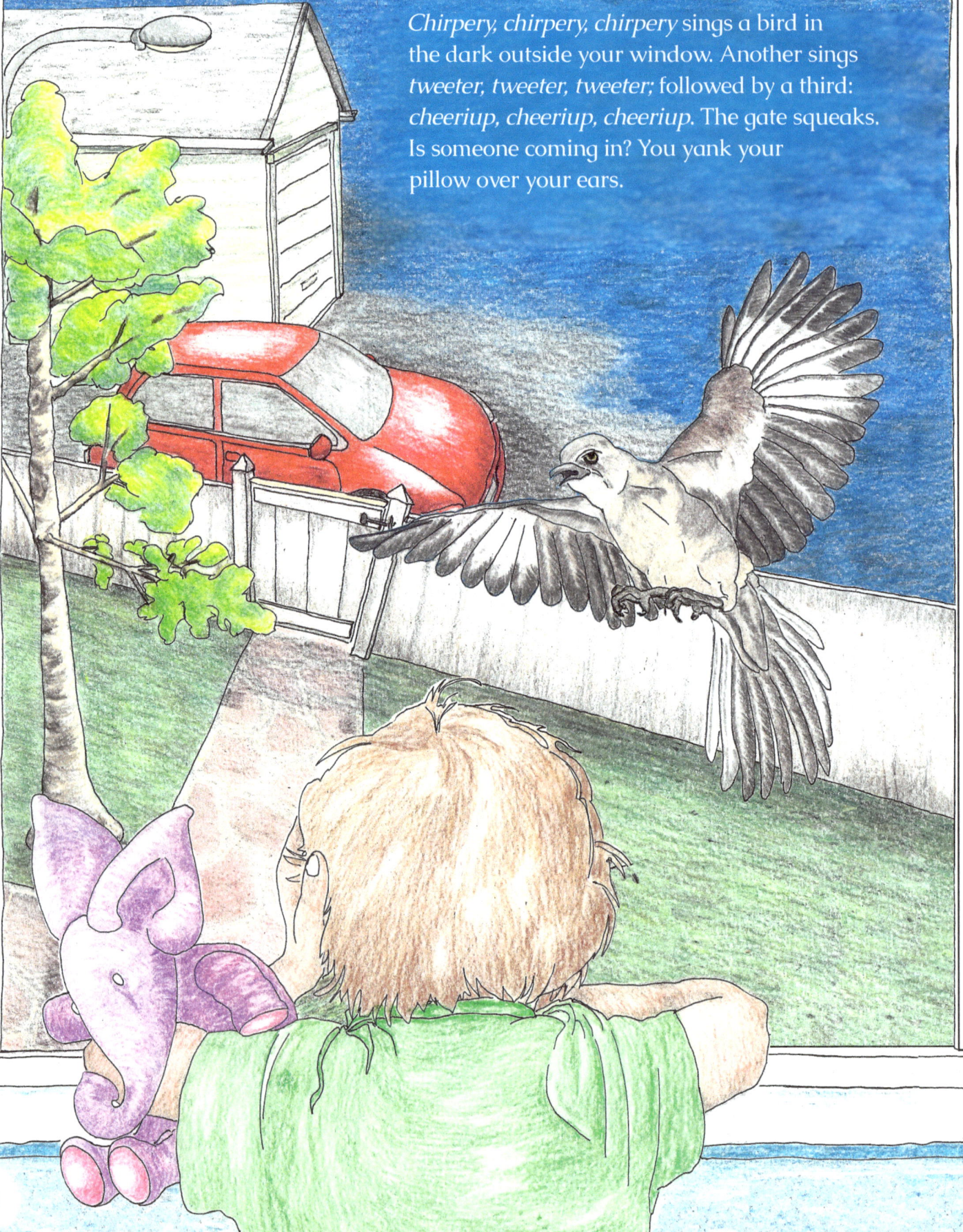

The next morning, the songs keep coming, from the tree right outside your window. There must be a dozen birds in that tree! A gray and white bird pops straight up from the treetop, sings *cheeriup, cheeriup, cheeriup,* flashes his white wing patches, and drops back among the leaves.

And the other noisemakers? The same bird flies to a telephone wire, perches, and sings: *chirpety, chirpety, chirpety, tweeter, tweeter, tweeter, birdee, birdee, birdee, squeakety, squeakety, squeakety.*

Hey! The squeaky gate came out of a bird's beak! One bird, and all those sounds.

Mockingbirds

are mimics.

Besides learning songs from their parents like any songbird, they copy sounds they hear—the car alarm, the squeaky gate, crickets, frogs, cell phones, other birds and animals. How many songs do you know? A male mockingbird learns from thirty to two hundred during his life.

Some people love mockingbirds. Some hate them, and all because a lonely male sings throughout a spring night, announcing his territory and advertising for a mate.

They're busy birds. Watch them, if you're lucky enough to live where they live, as they chase beetles and insects, which they eat along with earthworms and fruit.

Or observe two frenzied males facing each other across the invisible boundary separating their territories, hopping side to side or in circles, until one finally backs down and flies away. Their territory is important; it's their nursery, kitchen, and living and dining rooms.

Mockingbirds dive-bomb dogs and cats, like crows do. Do they have a sense of humor? Or are they warning dogs and cats not to intrude on their territory?

Song and Talk

Squawkers

Ahhh. You lean against the tree, closing your eyes after two hours of digging dandelions. A total ahhhhh!

D'wee d'wee...d'wee d'wee. Not moving, you squint open one eye and look up. *Deerily deerily...d'wee d'wee.* Too bad. Only feet and a white belly are visible through the leaves.

People like to see birds. But birds are often hidden. Some bird species look alike but sound different. Listening to bird song and talk (their vocalizations) is a great way to identify birds sight unseen.

Quackers

Which bird is vocalizing, and what is it saying?

True songbirds sing to declare their territory, and to advertise for a mate. They learn their songs by imitating their parents and other birds of their species. A baby robin in a cardinal's nest won't learn to sing like a cardinal—or a robin. (It'll be all messed up.) Birdsong isn't always "pretty" (think crows and jays), but it's always learned. Crows, cardinals, robins, sparrows, wrens, finches, and warblers are all songbirds. Depending on their species, they may learn only a few songs—or dozens, like the mockingbird.

Cheepers

Other birds' vocalizations are inherited, not learned. If you put a baby owl in a duck's nest, it will grow up hooting, not quacking. Birds like albatrosses, bitterns, eagles, and gulls are born knowing how to vocalize like their parents.

Twitter-ers

Squeakers

Birds don't just sing. They "talk" and call constantly. Go outside, close your eyes, and listen. You'll hear cheeps, chips, twitters—quiet contact calls as birds keep in touch while foraging for food. When signaling "danger," or when fighting for territory, the calls are loud and sharp. Baby birds peep endlessly for food, louder each day.

That's just the beginning. Think of the number of songs one bird has. Or the way its vocalizations change as it grows from nestling to adult. Or the different **dialects** sung by the same species who live in different areas.

Mockingbird? Mockingcrows are better!

Ever hear of a national dog-calling contest? No, but national bird-calling contests are big deals. A poet writing about the **dawn chorus** of cats? A violinist imitating a herd of gerbils? I don't think so.

Bird songs and calls have inspired artists forever, from every culture, and attracted many people to the outdoors— yes, kids too. It's all out there, waiting for you.

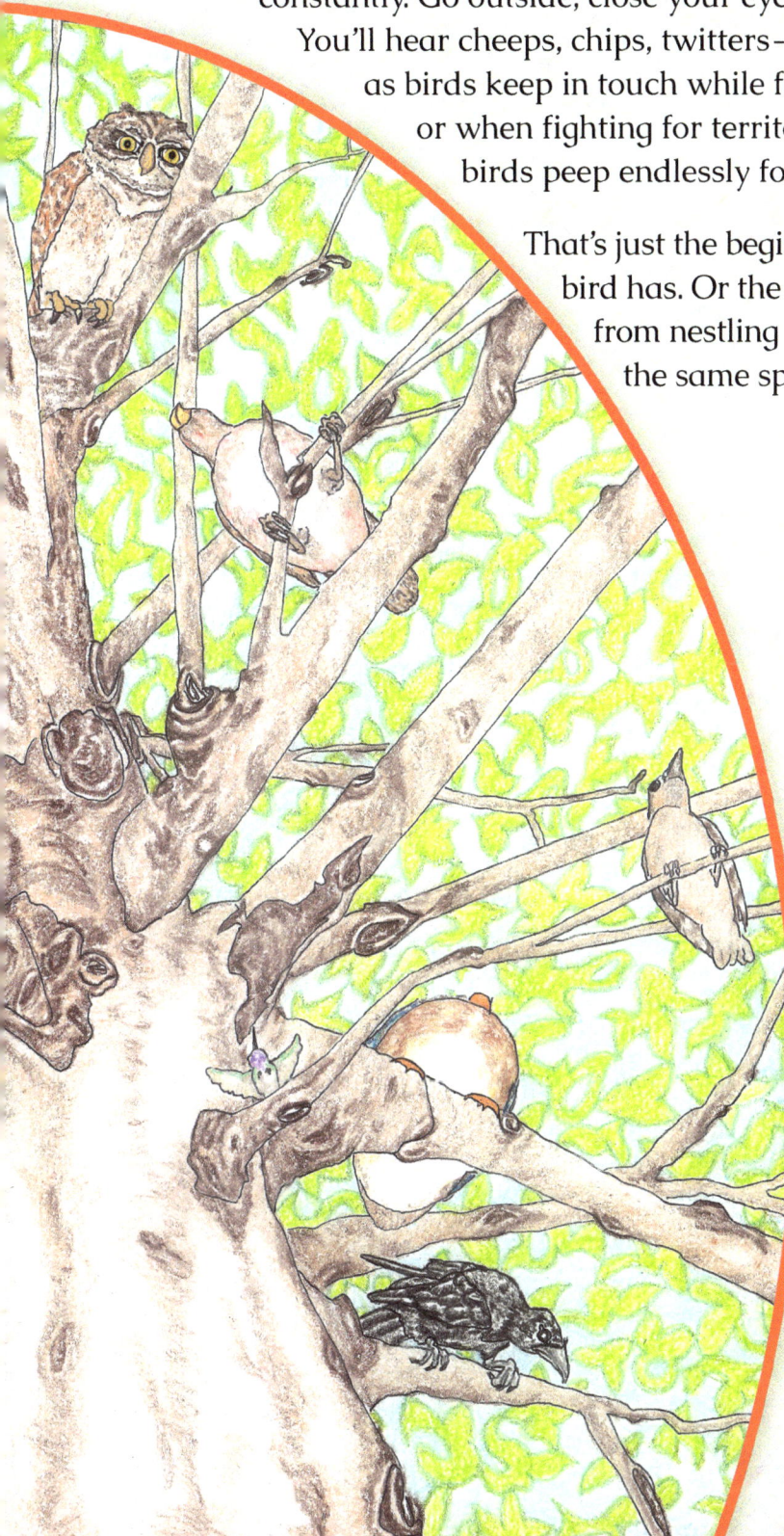

Hooters

37

You have relatives. Birds have relatives too. Hawaiian geese called

Nene

(pronounced "naynay")

are relatives of Canada geese.

Canada geese look like big ducks with long necks, and are very common all across North America. They like water; they swim with their goslings paddling in a line behind them. With their serrated beaks they graze like cows in fields and marshes on grass, seeds, and berries. They migrate in long V formations across the sky. But thanks to people and all our grassy fields and parks, some don't bother now. They can be pests.

Nene are smaller than Canada geese with longer, stronger legs and toes that stick out of their webbed feet, perfect for clambering on lava from Hawaii's volcanoes. For nene, swimming is fine but they prefer land, where they graze like Canada geese.

Migrating in long Vs? Not for nene, though they do fly among Hawaii's islands. Grasses? Definitely, but they add Hawaiian specialties such as blossoms, and 'ōhelo and pūkiawe berries.

Nene almost became extinct, like dinosaurs—no more nene, ever again. The reason was—people.

At first there were thousands of nene and no people on the Hawaiian Islands. People came in boats with their cats and dogs and pigs, and the rats that snuck aboard. To control the rats, Hawaiians imported a small animal from Asia called a mongoose (which looks like a cat, not a goose) to eat them. Unfortunately, the mongooses—as well as the rats, dogs, cats, pigs, and people—ate the nene eggs and babies. Soon there were only thirty nene left in Hawaii, which meant in the entire world.

But Hawaiians liked their nene. In 1957 they were made the **state bird** of Hawaii. Hawaiians quit hunting them, and worked hard to control the rats, mongooses, wild dogs, cats, and pigs. They caught the nene and kept them safe while they laid eggs and raised their goslings, and then they freed them. Slowly, slowly, the number of nene has increased, to several thousand! You can see them on the volcanoes, and, like Canada geese, on fields and golf courses.

Volcanoes = Bad
People = Worse!

Are **Owls** magic?
Mysterious? Scary?
People used to think so.
Why?

For one thing, owls have flat faces like ours, and two round staring eyes. Most other birds look at you with one eye at a time. Watch them.

Most owl species are **nocturnal**; our night is many owls' day. They fly silently, like ghosts, because of their special wing feathers. They have eerie calls: hoots, hisses, screeches, screams, coos, clicks, chitters, whistles, and wails. They even clack their beaks and clap their wings. Scary at night, when you can't see them!

Owls are not vegetarians! They gulp their prey down head first, or slice them up with their snapping, clacking bills: birds and eggs, rodents, weasels, skunks, snakes, frogs and fish, worms, caterpillars, moths, insects— anything catchable, even house cats. They bring up the indigestible parts in pellets like eagles and gulls do.

Owls' flat faces concentrate and magnify sound. They hear so well they pounce on mice running in tunnels underneath the snow. Their nocturnal vision is so good they swoop down on prey unfortunate enough to be out after dark.

Here's something strange: owls don't make their own nests. Little owl species (one is smaller than a sparrow) nest in old woodpecker holes in trees or cacti, or prairie dog tunnels. Big owls (one is almost as big as an eagle) reuse crow or squirrel nests, or find a flat ledge in a barn or on a cliff, or they nest on the ground.

What people find magical or frightening about owls are evolutionary adaptations, helping owls find food, a roost, a mate, and someplace to nest. Our nineteen North American species live almost everywhere: forests, meadows, prairies, deserts, alpine tundra, cliffs, mountains, golf courses, farms, and towns.

Wow!
It's a corvid
flash mob!

Other birds are scared of owls. If they find
one camouflaged in a tree trying to get a good
day's sleep, they mob it! All the neighborhood
birds, even hummingbirds, dive-bomb it,
cawing and cheeping and screeching until the
poor owl gives up and flies off to find another
roost.

Because most owls are active at night and
hidden by day, they are hard to find. But
listen for them at dusk and after dark. As
silently as an owl flies, follow their hoots and
screeches. When you think you're close, turn
on your flashlight and aim it at the sounds.

Feathers

It's a winter day. You put on your coat, hat, and gloves, and head outdoors.

Spring comes, and rain. Time for boots, a rain jacket, and 'brelly.

Then summer, and you spend most of the day in your swimsuit. And, you dress up for a party!

Birds do all those things—with feathers.

In the whole world, only birds have feathers.

Feathers keep birds dry, protected, and insulated so they're neither too warm nor cold. Dull feather colors camouflage birds. Colorful feathers help them attract mates. Without feathers, birds couldn't even fly.

Feathers are like our hair; after they grow out of the skin, they are dead. They wear out, and once or twice a year the old feathers are pushed out by new feathers. That's called molting.

Like hair, feathers don't fall out at the same time. (That would make a very naked, unhappy bird.) Wing and tail feathers are replaced two by two so birds can fly while they're molting them—except remember the ducks?

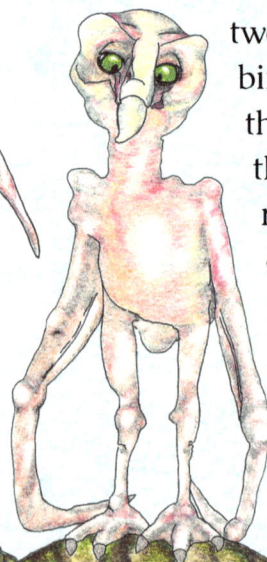

42

Birds look like they have feathers all over their bodies, right? But most of birds' feathers grow in rows called feather tracts, alternating with naked spots (covered by feathers). That way a bird doesn't need to make or carry around so many feathers. (For birds, feathers aren't lightweight—all together they weigh more than the bird's bones.)

Birds have different kinds of feathers:

Hundreds of *contour feathers* cover birds, and give them their normal birdy shape and colors. Contour feathers, which include the flight feathers, have downy fluff at the base, to help keep birds insulated.

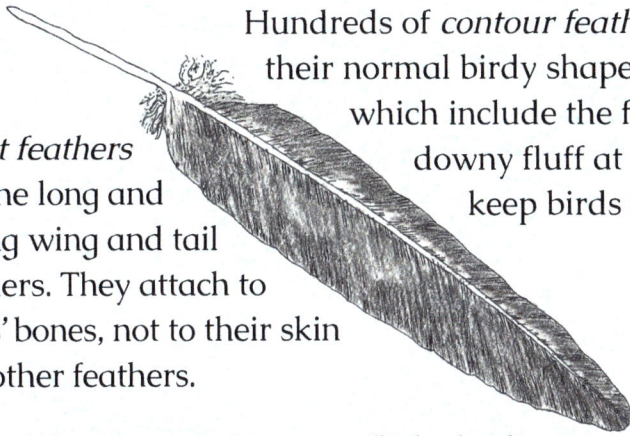

Flight feathers are the long and strong wing and tail feathers. They attach to birds' bones, not to their skin like other feathers.

If your arm were a wing, your flight feathers would attach from elbow to fingertips. (The rest of your arm would be covered with regular contour feathers.) Tail feathers connect to the bird's tailbone.

Many birds, especially water birds and northern birds, also have *down feathers*, soft, fluffy feathers hidden under the contour feathers for extra warmth.

Tiny muscles attach to feathers, so they can move—to help steer (flight feathers) or fluff up for warmth (contour feathers).

Look at a feather. There's a central *shaft*. *Barbs* come out on each side of the shaft. (The barbs on a flight feather are **asymmetric**—they're shorter on one side.) You can't see the next tinier part of the feather, but each barb has hooked *barbules*. The tiny hooks grasp one another, connecting the feathers together into one smooth **aerodynamic** surface—insulated, waterproof, windproof, and streamlined.

More amazing feathers-learning in eXtinct! (Page 65)

shaft

barb

barbule

43

Most fish-eating birds catch fish one by one. Not **Pelicans**

—they catch them by the dozens in their own "fishnet," a pouch attached to the bottom of their bill.

People used to think pelicans stored fish in their pouches, like you'd put food in the refrigerator. Actually, the gular pouch, as it is called, is used to catch fish—along with a lot of water. The pelican strains out the water and, tipping its head up, swallows the fish whole, and the pouch looks like a bill again.

There are two species of pelicans in North America, and they catch fish in two different ways:

White pelicans, mostly freshwater birds, fish **cooperatively**. They swim in a line, flapping their humongous wings (eight feet, wingtip to wingtip) to herd fish into shallow water. Then they scoop them up.

Brown pelicans, saltwater (marine) birds, fly above the water looking for schools of fish. When they see them, they plunge straight down, closed bills slicing through the water, heavy bodies hitting with a huge splash, to capture them. And they have a new technique: begging fishy treats from fisherfolk!

Pelicans live in colonies, roosting and nesting together like flamingos. They fly in lines, flapping their wings or gliding in unison.

Pelicans don't sit on their eggs. Like penguins, they incubate them with their feet. And of course, they don't have incubation patches. (Why not?) Pelican chicks are altricial and dependent on their parents for several weeks. Then they're off to pelican day care, the crèche.

The firstborn chick usually abuses and kills the younger, smaller **sibling**. (That's called siblicide.) The second chick is a "backup" in case the first doesn't make it.

In many other species, like gulls, owls, and eagles, siblicide is common too. In nature, the survival of the whole species is more important than the survival of any one bird, and a single chick is easier to raise if food is scarce.

You're supposed to chew each bite twenty times, you know.

Aren't younger human siblings fortunate?

45

You hear Quail

more easily than you see them.

One species whistles *Bob Whi-i-ite*. (And the name is...?)

One species calls, *Chi-CAG-o*, (like the city). But it's named the California quail, because it lives in the West.

Quail are robin-sized birds, but weigh considerably more, and have short tails and fancy head feathers. California quail have bobbing **topknots**, and male bobwhites have crests.

Their ground-colored feathers camouflage them from their many hungry predators— hawks, coyotes, snakes, weasels, raccoons, opossums, foxes, and **feral cats**—and human hunters! They are prey birds.

Unless they're nesting, quail live in flocks called coveys—so usually you see more than one quail. (Oddly, the plural of quail can be quail or quails.)

Quail are scritch-scratch birds. They kick aside leaves and dirt to find seeds and insects. Their preferred **habitat** is weedy, seedy, shrubby open country, from deserts and forest **edges** to your own neighborhoods.

While the covey searches for food, a male acts as a guard (sentinel), who calls "danger" if it sees a predator. The quail scatter in all directions. They would rather walk or run than fly, but sometimes they all burst into the air at once. That startles the predator! When it's safe to gather again, the sentinel calls the "all safe" signal.

Most quail roost together in trees or bushes. Bobwhites roost on the ground in a circle, all facing out, to share their body warmth and be on guard for predators.

During nesting season, the coveys separate into families. The female lays a dozen eggs in a ground nest, and likely will have two batches—a couple dozen chicks per breeding season, which helps maintain the population in spite of predators! The babies leave their ground nest soon after hatching; they're precocial, like ducklings. After nesting is over, quail return to their coveys.

Remember, quail are flocking birds. When you see one, look for more!

It says here that hunting fees provide about a billion dollars for **conservation** every year.

I'm just glad those humans don't eat crow.

Close your eyes and picture an American Robin.

They live throughout North America, from Alaska and Canada to Central America, from mountains and wilderness to cities. It's easy to find and observe them wherever you live.

Many robins migrate south for the winter. There you'll see flocks of them eating fruit, bugs, and worms. If you live in the north, one winter day you'll see robins heading your way, singing, *Cheer up, cheerily, cheer up, cheerily, cheer up!* Then you'll cheer up, because even if it's still snowy and cold, you'll know spring is coming.

People say, "The early bird gets the worm." That's the robin. They start the dawn chorus each morning, and they love earthworms. Watch a robin hop, stop, cock its head, and pull a worm from the ground. Robins seem to be listening to find worms, but they're using their eyes—earthworms don't make much noise. They tilt their heads to see a worm, since birds can't roll their eyes around like we can. (Remember bitterns?)

In the spring, if you see a robin with its mouth full of twigs or mud, watch it, because it is building its cup nest in a tree or bush, or maybe under the eaves of your house. Robin parents feed their nestlings a couple hundred times a day! After the babies fledge, the mother starts another family, and the father feeds the fledglings. He follows them around, popping worms and insects into their wide-open mouths. Then they follow him around, begging. Finally, he drops the worm on the ground, and they learn to feed themselves. Fledglings look like their parents, except for their spotted breasts—and their begging.

You mean humans have to buy oil?

Robins love baths. They crowd around a puddle or birdbath taking turns, chirping, and splashing. Then they fly to a tree, fluff out their feathers, and comb (**preen**) them with their beak.

All birds preen to keep their feathers healthy. Without healthy feathers a bird cannot fly, or stay warm and dry. At the base of its tail, a bird has something called a preen gland, full of feather oil, like built-in hand lotion.

Birds' necks are extremely flexible to allow them to preen with their beaks. (No arms and hands!) They rub their heads and beaks in the oil, then run each feather through it. They use their claws to preen their head feathers.

Robins are one of our favorite birds. So it's really sad that many fledglings are caught and eaten by house cats before they learn how to fly well. It's much safer for birds if cats stay indoors. (It's much safer for the cats, too. In some areas, coyotes' diets are 60 percent house cats.)

"Dad, I see smoke going DOWN a chimney!"

The girl's father hands her a pair of binoculars. She focuses and sees that the "smoke" is thousands of migrating

Swifts,

swirling above the chimney, and then diving right into it.

Chimney swifts used to roost in hollow trees. People chopped down the trees, but they built chimneys, and swifts have adapted. Someone counted twenty thousand swifts flying into one chimney—their migration motel.

CROW MOTEL
VACANCY

You want a room for how many?

Except for roosting and nesting, swifts spend all of their time in the air. They swoop over water, catching a drink, and a splash-bath. They can't perch at all; they roost upright, clinging to the side of something, like a cliff—or chimney. Many species sleep in flight.

Swifts grab grass, sticks, and mud on the fly to make their nests, which they glue onto a cliff or inside a chimney with their spit (saliva). They collect hundreds of insects in a spitty ball to feed their babies.

"Look! The Swallows arrived today!"

The darting, dainty swallows swoop over the pond as if they surely remember that very pond from the previous year.

Swallows migrate north, catching insects as they fly (like eating while you drive). Their arrival dates differ each spring, depending on when insects hatch.

Like swifts, their food flies, and swallows don't spend time on the ground. Their small feet fit perfectly around telephone wires though, and sometimes you see dozens perched, twittering.

People and swallows get along. Swallows devour millions of pesky insects. They love flies! They often nest around people. Barn swallows (the swallows with the forked tails) attach their mud and grass nests up under the eves of buildings. Other species nest in tree cavities, birdhouses, or tunnels in riverbanks. In the East, purple martins use bird "apartment houses" people make just for them.

Is it a swift or a swallow? You peer through your binoculars, trying to follow the bird's darting flight.

They can look similar, though they are not relatives. If they are perched, they are swallows. Swallows are more colorful, and their flight is more swooping. Swifts, more closely related to hummingbirds than to swallows, are darker and fly higher, flapping and soaring, making lightning turns in pursuit of insects. Their wings curve back; some people call them "cigars with wings."

Go on, go outside.
Is it spring yet?

Migration

Oh, the stories people concocted to explain how their birds disappeared every winter!

Birds hibernated (slept through the winter). They flew to the moon. They hitched a ride on bigger birds. They turned into winter birds. The barnacle goose supposedly became a barnacle for the winter!

Actually, birds survive winter in three ways: they *hibernate, tolerate, or migrate.*

Only the *common poorwill* hibernates. It enters a state of deep sleep, lowering its body temperature and heart rate for days or ,weeks. Other species like swifts and hummingbirds may drop their heart and breathing rate and temperature to survive cold nights. (That's called torpor.)

Some birds tolerate winter, because their winter is warm, or because they've adapted to cold winters. (What are your winter birds?)

Half our bird species (billions of birds!) migrate. Most fly from their summer home, where they raise their chicks, to a winter home farther south. Some mountain birds just migrate lower to warmer winter weather. Others follow seed crops.

Why do birds migrate?

Most birds fly north in spring to avoid the crowds of nesting birds in their winter home, and to snatch up billions of northern insects to feed their babies. They fly south as the days shorten and grow colder, and food — insects and seeds — is harder to find.

Where do North American birds go?

They summer in the United States, Canada, and the Arctic. Winter it's mostly the southern United States and Central and South America.

How do birds do it?

They prepare. They molt new, strong flying feathers. They gorge (eat enormously) and store fat for energy. Body parts they don't need for migration shrink. Many birds fly at night when it's cooler, and stop to eat during the day. Insect eaters catch their food on the fly. Raptors ride **thermals** (drafts of rising hot air) and hardly have to work to fly.

How do they know when to go?

Birds are born with an instinctive calendar that tells them to go when days are shorter in the fall and longer in spring.

How do birds find their way?

- Some juveniles (kids) migrate with their parents and learn the route.

- Others find their way without their parents, by instinct.

- Some birds navigate like ancient sailors did, using the position of the sun (daytime) and stars (nighttime).

- Some species have tiny magnets inside their brains, like **compasses**.

- As they get closer to home, they recognize landmarks that guide them to the same neighborhood they left months before.

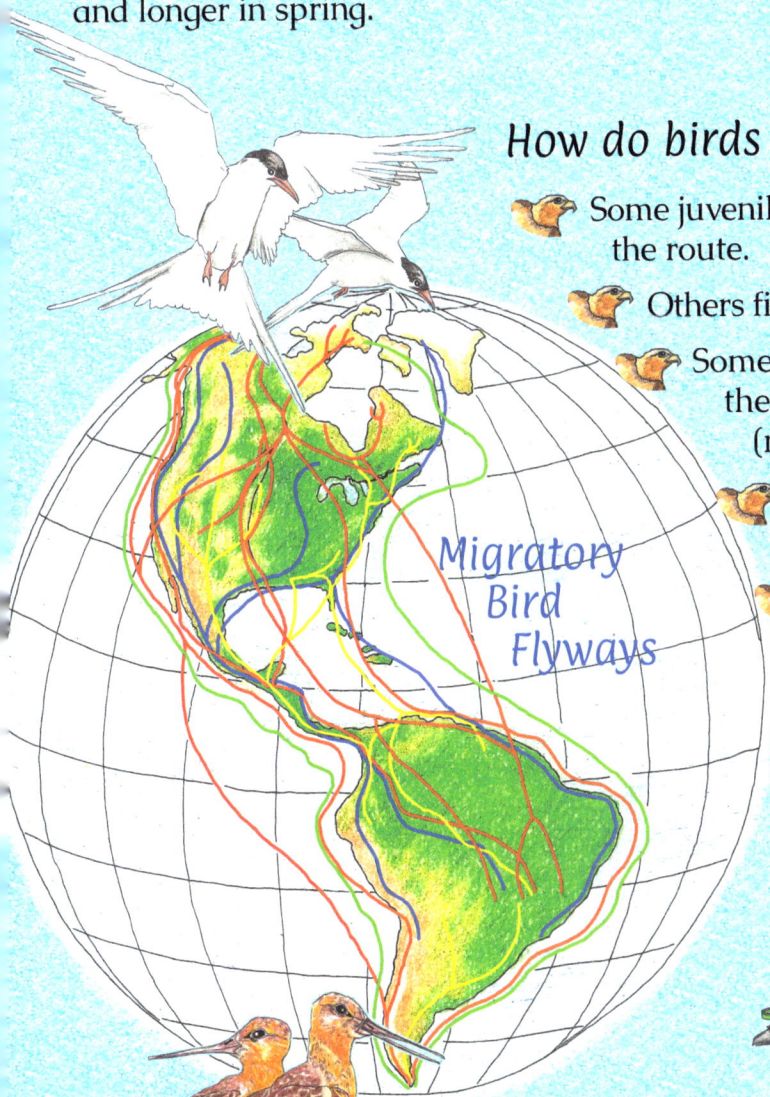

Migratory Bird Flyways

Compass says turn right at the next island.

Migration miracles:

Bar-headed geese fly over Mount Everest.
Bar-tailed godwits can fly 7,000 miles nonstop.
Snow geese migrate in flocks of over one million.
Tiny hummingbirds fly hundreds of miles in one night.
Northern wheatears, weighing less than one ounce, fly 9,000 miles.
Arctic Terns fly 44,000 miles, round-trip, from Antarctica to the Arctic.

Animals as large as whales and as small as butterflies also migrate. But birds have adapted to use more territory than any other animal on earth, partly through migration.

On Thanksgiving the Turkeys

we gobble up are mostly domestic—not wild, but raised on farms.

Yet just about everywhere in North America, you'll find wild turkeys, constantly turkey-talking. Like ducklings, baby turkeys (poults) peep to each other and their mother before they hatch. At just two days old, they know her by her voice. They learn her commands, like, "Enemy nearby, freeze!" or "Run!" or "Come to me!"

Turkeys live in flocks, and watch out for one another. If they find some delectable seeds, they have a special call to let the others know. Like quail, they have a "danger call." They even have a call for when it's time to hop out of their roosting trees each morning.

And the big male turkey (called a Tom) looking for a female? "Gobble, gobble!" he says, throwing his bald head back, showing off his red wattle (comb), fanning his tail, and dragging his wings on the ground. To a female turkey, he is a handsome fellow!

Every turkey—young, old, male, or female—knows its power place in the flock. That's called the pecking order, because turkeys (and other birds) fight by pecking, kicking, and hitting with their wings to learn who among them is top dog. Once a turkey has learned its place, there's less squabbling. And yes, dogs and other animals, including people, want to know where they rate in their pecking order too.

Turkeys eat seeds and nuts, fruit, grass, ferns, even cactus, insects, salamanders—about anything of the right size. As a turkey swallows, some of the food is stored in a special pouch in the esophagus (the tube between the mouth and the stomach) called the **crop**—like a kitchen cupboard.

Eat your grit. Mama Turkey says it's good for you.

Yech!

Turkeys have no teeth, right? Well! They swallow everything whole. (Maybe that's why, when we chow down, people say we "gobble" our food!)

They also eat pebbles or dirt, called grit. Part of their stomach called the gizzard is a big grinder, and it uses the grit to help grind up all those seeds and acorns. Turkeys don't need to cough up pellets like owls and gulls do. (Why not?)

Ingenious! Gizzard plus grit equals teeth.
Other seed-eating birds have crops
and gizzards, too.

"Urban" means a city or town.

Urban birds

live, breed in, or migrate through cities.

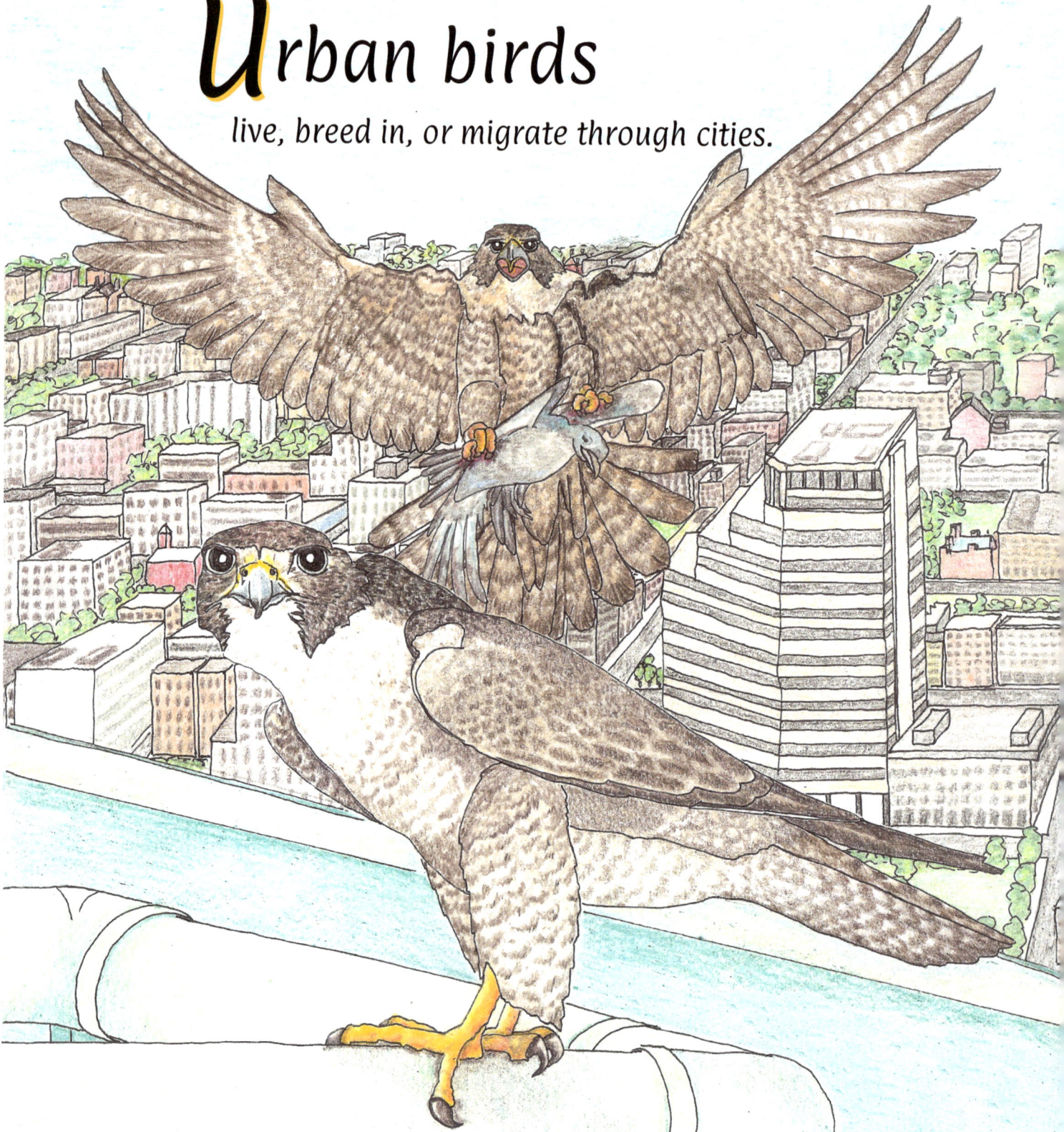

People think of cities as unending concrete, honking cars, and tall buildings. But get this: over three hundred species of birds have been seen in New York City. One hundred species even nest there. How?

Think of the city's habitats; besides all the concrete, there are ocean beaches, estuaries (where ocean tide meets river), rivers, streams and ponds, forests, shrubs, parks, lawns—and feeders, bird baths, and trash.

Ducks find ponds, and bitterns find marshes. Songbirds thrive in Central Park, where jays call raucously from tall trees. House sparrows search for seeds and breadcrumbs on sidewalks. Pigeons and crows find garbage and window ledges and tall statues.

One special bird is the *peregrine falcon,* a raptor (bird of prey) that eats other birds. Though cliffs and open country are its usual habitat, it has adapted to New York City because it never runs out of food there— pigeons and starlings for starters.

Another raptor, the *red-tailed hawk,* also lives there—rats, yum! Instead of cliffs, falcons and hawks nest high off the ground on bridge supports, balconies, and window ledges.

Migration? Millions of birds migrate right through New York City, the largest city in the USA. Likely their ancestors migrated there before there was a city.

The migrants use ancient flyways (bird highways), which have never changed.

Because of the varied habitats, the migrants drop out of the sky to eat and rest before flying on.

But cities are hard on migratory birds. City lights confuse nocturnal migrators. Birds crash into skyscraper windows, are hit by cars, or caught by cats, dogs, and coyotes.

Think of the city's three hundred species— three-fourths of all of the species seen in New York state, a quarter of ALL North American species—who have adapted to an urban life so very different from their original habitats!

The turkey *Vulture* (misnamed a **buzzard**) circles,

wings outstretched, the hot wind riffling her feathers. Her body rocks in the air. She smells something good. She flaps her wings once, circles again, and sees the dead cow beneath her.

As she floats in for a landing, she startles two coyotes that got there first.

The vulture lands with a small "thump," hopping sideways a couple of times, and waits. Finally the coyotes, their stomachs bulging, trot away. At the entrance to their den, their pups run up, whining and nipping at their parents' mouths. The parents regurgitate dead cow, and the babies eat.

The vulture hops to the cow's body. Her beak can't pierce tough cowhide, but the coyotes have done that for her. And if the carcass isn't opened up? The vulture pecks out the eyeballs, her favorite, and then eats the tongue or other tender parts she can get to.

Another vulture lands. The first vulture hisses and raises her wings (pecking order), then sticks her bald red head deep inside the cow's belly, while the other waits.

Vultures' heads are bare, so they don't get their feathers icky from the blood and guts.

When the vulture finishes, she leans forward, stumbling and flapping her wings until she is finally airborne. She flies to a nearby cliff, and disappears into a narrow crack.

Inside the cave, she clucks a couple of times. Two chicks hop over to her, hissing, stomping their feet, and jabbing at each other (pecking order again). When they reach her, she regurgitates dead cow into their beaks.

If you didn't have hands, you might carry food to your babies in your mouth (beak) or throat like songbirds, or stomach, like the coyote, albatross, and vulture.

Vultures join scavengers from beetles to bears in disposing of carrion, which makes our earth cleaner and safer. They eat rotting meat just fine. They can be poisoned, though, by the various human **toxins** in the carrion, such as **pesticides**. Then there's **lead shot**, from animals killed but left by hunters. Some states (and countries) have passed laws to prevent hunters from using lead shot, which is fortunate for our crucial scavengers.

Isn't it weird? A bald eagle isn't bald, but a turkey vulture is. And, of course, it's not a turkey. It should be called a bald vulture.

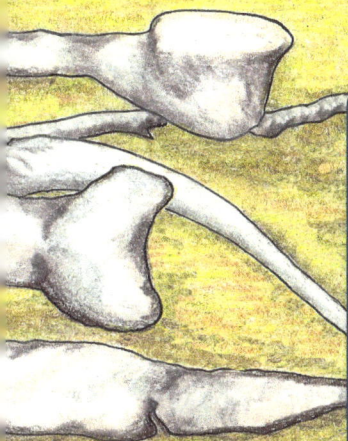

Whatcha gonna have?

Blue Bird Chef's Special

Tongue
Eyeballs
Stomach fraîch

Ask about our
Tasty Tongue Treats

Woodpeckers
peck wood.

Obvious, but there's more to it. Trees are as important to woodpeckers as water is to ducks.

Woodpeckers don't usually perch like songbirds and raptors. They cling upright to tree trunks, using their strong, stiff tail feathers for support and balance.

Trees mean food.
Woodpeckers eat nuts and acorns, bugs, and caterpillars. They look for insects on tree bark, and explore woody crannies with their very long sticky tongues. They hear bugs inside the tree, and excavate (dig or peck) to find them.

Woodpeckers called *sapsuckers* drill neat rows of holes into the tree bark. Tree sap (which is like the blood of the tree) oozes out. Bugs get stuck in the sticky sap. Sapsuckers drink the sap, and eat the bugs and the inner bark of the tree.

An exception: a woodpecker called a *flicker* eats ants—but usually from the ground! Anthills are a special favorite.

60

Trees mean family.

With their strong beaks, woodpeckers chip out cavities for their nests.

Trees mean sleep.

Woodpeckers roost in their tree cavities, sleeping upright like a swift. If a predator like a snake or weasel comes in one hole, the woodpecker exits from another.

Trees mean communication (called drumming).

Woodpeckers drum by striking their beaks loudly and very rapidly on a hollow tree, or sometimes a metal chimney, or—oops—the side of a house. (It's a woodpecker drumroll.) They're announcing to other woodpeckers, "This is my territory," or "I'm looking for a partner." Their bug-hunting or excavating taps are quieter, slower, and more irregular.

Woodpeckers also have calls, but they are not songbirds, and do not sing.

Wow! It's 2" and has a sticky end!

Best Drummer Award

The Birds

Other species are dependent on woodpeckers. Owls, wrens, and small ducks like bufflehead nest in old woodpecker cavities. Other birds and animals eat insects and sap from sapsucker holes.

Most woodpeckers are black and white, spotted or striped, so they blend in with tree trunks. Flickers have ground-colored, brown-striped backs: camouflage again. Males are more likely to have red on their heads.

How do you fit 5" (13 cm) of tongue back in a tiny head?

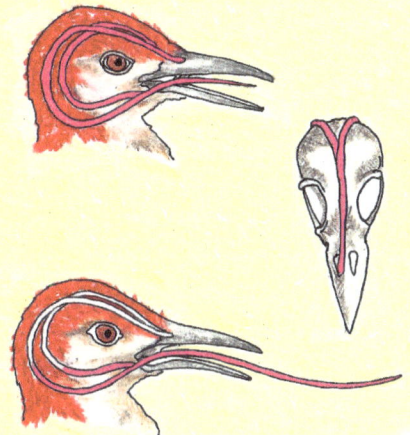

Reproduction

Every living species must reproduce—that is, make more of itself—to avoid extinction. For birds, reproduction includes establishing a territory, courting (like dating), nesting, mating, incubation, and raising the chicks until they can live on their own.

Birds start making babies in the spring. They court to win a partner, with flashy colors and birdsong. (Lucky us!)

But courting isn't just song and bright feathers. It's also behavior.

Males find a safe territory with a good food supply, and show it off to prospective females. They feed them, or build model nests. Like the eagles, some birds have dazzling **display flights** or dances, often including the females. All the males' tricks exist to convince the females that they're strong, healthy, and good providers. Usually the females get to choose their mates.

The relationship between mates is called a pair bond. How long does it last? For some species, like hummingbirds, it's as brief as courting and mating. For others, it lasts for one or more nesting seasons, even for life, and the parents raise their offspring together.

After the pair bond is established the birds mate, which produces **fertilized eggs** and also strengthens the birds' pair bond. Their bodies join briefly, because making a baby involves combining **genes** from both the male and female. Genes are nature's **blueprints** for creating another bird. The male's genes are in his sperm, made in his testes. The female's are on her egg **yolks**, made and stored in her ovary.

Birds have one exit-hole, the cloaca, for mixed urine (pee) and feces (poop), and for sperm (to go up) and eggs (to come down for laying).

When birds mate, the male hops on top of the female. She pushes her tail aside so their cloacas touch. Millions of sperm enter her cloaca, and "swim" up a special tube toward her ovary. One sperm enters the germinal spot on the yolk, and the parents' genes combine right there, creating a fertilized egg! Mating takes just a few seconds; mating birds are easy targets for predators. As the fertilized yolk travels down to her cloaca over a day or so, it receives the "egg white" which nourishes and protects the **embryo**, then the shell. Then the egg is ready for laying.

Inside an egg is everything necessary to make a new bird!

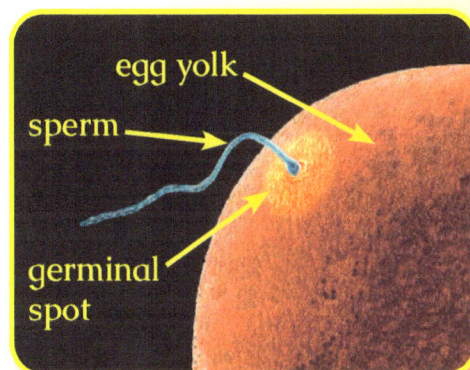

egg yolk

sperm

germinal spot

Tell me. Are we really gonna do this again next year?

MeeMeeMeeee!

Nooo MEEEEEEEE!

Meeeeeee!

63

"eXtinct"

means gone forever...

Archaeopteryx, a dinosaur-like bird or bird-like dinosaur, lived 140 million years ago. It had feathers, wings, teeth, and a long bony tail. It and other bird species went extinct long before people existed.

But in the last couple hundred years, we humans have caused the extinction of dozens of bird species. (Remember the nene?) To make room for more people, birds' habitats were destroyed. Birds were hunted and their eggs collected for food or fun. Human-made toxins harmed them.

Scientists shot birds to study them!

Carolina parakeets flew in noisy flocks around the southern US, perching on winter trees like Christmas ornaments. After farmers cut down the trees, parakeets ate their crops. So farmers shot them. Our parakeets were gone by 1913.

Two hundred years ago, five *billion* passenger pigeons thundered overhead, darkening the sky during migration. People shot them by the millions. (There were special pigeon cookbooks.) "Martha," the last of those five billion pigeons, died in a zoo in 1914.

Huge *ivory-billed woodpeckers* lived in forest swamplands in the southern US. Logging and farming destroyed their habitat. Settlers killed them for food and their amazing bills.

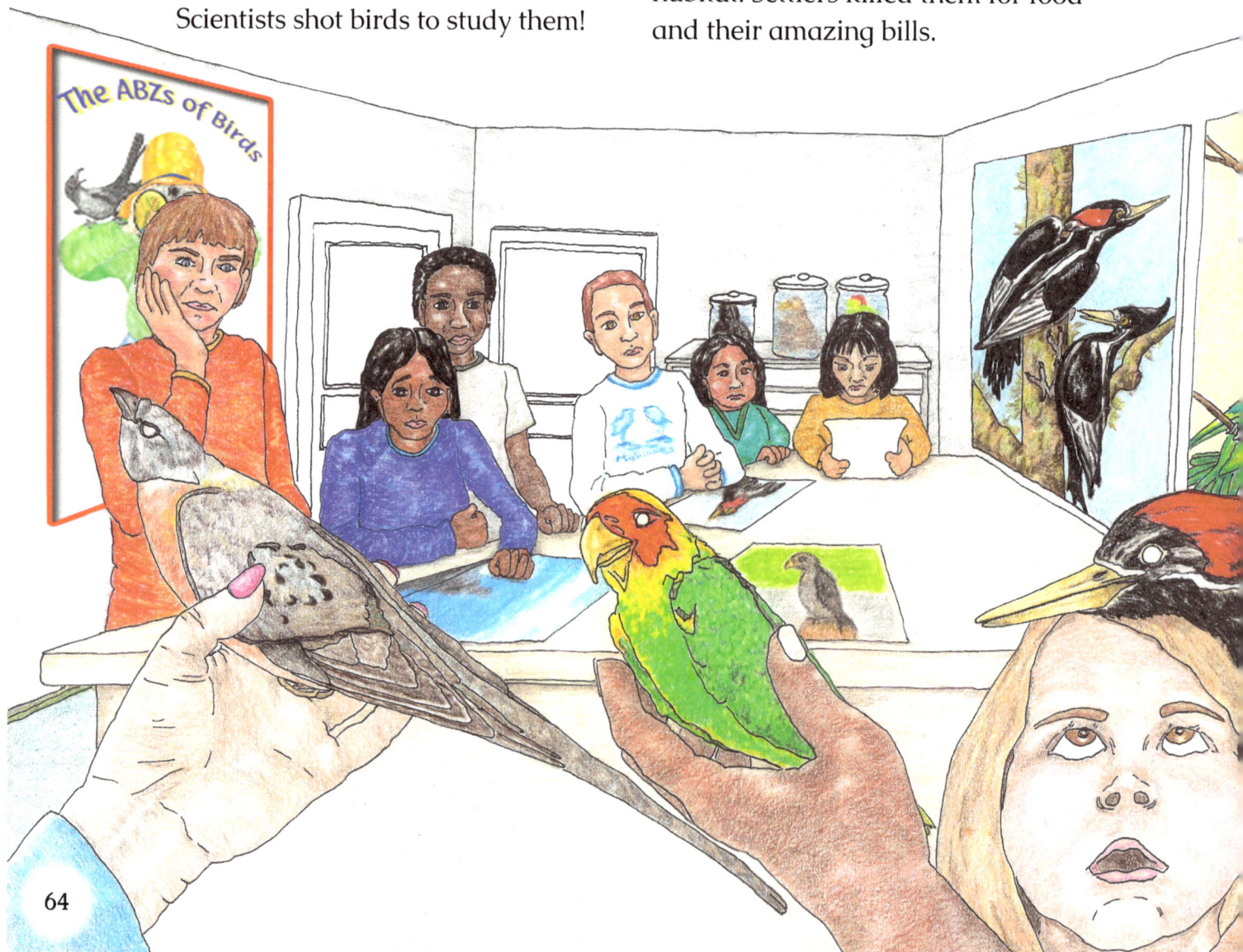

By the 1950s they were extinct…or were they? Some people think they have seen one or two in remote southern swamps. If so, will we kill them, too?

Guess who started the conservation movement to protect birds? Two women! In the 1890s, they learned that five million birds were killed annually to decorate women's hats. Horrified, they arranged tea parties to dissuade women from wearing dead birds. The movement grew, leading to creation of the National Audubon Society, and now many other conservation groups. Laws were passed to protect birds. Scientists now use binoculars, cameras, bird **banding, radio transmitters**, and radar instead of guns to study birds.

There have been amazing victories. Birds of prey became **endangered** because of DDT use. DDT is a pesticide that softens egg shells, killing the chicks. Fortunately, due in part to **Rachel Carson's** book *Silent Spring* (as in no birds singing), the US banned DDT in 1972, and bird populations began to recover. Unfortunately, DDT is still used elsewhere in the world, and other very toxic pesticides, some of which are also endangering our bees, are used here and everywhere.

Plain black feathers are a good thing!

The struggle never ends. Remember though, a few people saved millions of birds. Think we can?

PASSENGER PIGEONS

AR 694110 1902 FORT HENDERSON ARK

Harriet Hemenway Minna Hall

People call **Yellow warblers** "butterfly birds."

They're bright yellow, and flit about in shrubs and small trees, usually around wetlands, searching for insects.

Like many bird species, they like edges, where different habitats like streams, forests, gardens, or orchards meet.

There are over fifty species of warblers in North America. Though none are as bright as the yellow warbler, most species have at least some yellow feathers. They're insect eaters, live in trees and shrubs, and migrate.

Warblers nest and raise their families with us in North America. In the fall most migrate to the southern US, Central and South America. They fly nonstop all night. During the day they land, rest, and gorge.

People who live in Florida, Louisiana, and Texas celebrate spring when they see the first exhausted, famished warblers (the **migration fallout**) dropping out of the skies by the hundreds on their way north. Their nonstop flight over the Gulf of Mexico can last twelve to twenty-four hours, depending on the weather.

Each day they eat over half of what they weigh. That would be like a fifty-pound kid eating (gasp!) 450 hot dogs a day!

Perfect! Her weight in insects means she gets half and I get half.

One of the warblers' biggest enemies is a small brown bird with a high squeaky voice, the *brown-headed cowbird*. Cowbirds are brood parasites, which means they don't make their own nests, incubate their eggs, or raise their babies. They sneak their eggs into another bird's nest, for example, a warbler's.

The cowbird hatches first, so it's bigger and stronger than the warbler nestlings and demands food more aggressively. If the warbler parents don't catch on, they feed the cowbird chick preferentially, even if it means their own young don't get enough to eat.

If the cowbird host (in this case the warbler) realizes that one of the eggs isn't her own, she'll build another nest right on top of the old one, and lay another set of eggs. Or the warblers will renest somewhere else, trying to outwit that cowbird, who in turn keeps trying to trick the warblers into raising her babies.

Zebra finches

are small, grayish birds with black and white stripes.

You can see how they got their name.

In North America, they aren't wild. They're pets.

Pet birds are totally dependent on their human owners and need daily care. Zebra finches like roomy flight cages, and please, a fresh bath every couple of days. They eat grass seeds, fruit, and greens. They must be given cuttlebone (high in minerals like calcium) to keep their bones strong. (Cuttlebones are bones from cuttlefish, which aren't fish, but a relative of the squid.) When the finches' owner wants babies—and assuming he has a male and female—he puts a basket and nesting material in the cage.

Zebra finches were imported from Australia, where they live in large flocks in the grasslands. There, they eat grass seed and tiny insects. Of course these zebra finches have never seen a cuttlebone! They get everything they need from their habitat. They nest after rain, when plants bloom and produce seeds. If the weather becomes unpleasant in one place, they migrate to another part of Australia.

Too many zebra finches were being caught and exported for pets. In 1960, Australia's government said, *Stop!* Now our zebra finches are the offspring of pet finches and likely couldn't survive life in Australia's grasslands.

The fate of the zebra finches is like that of other native bird species around the world whose numbers were being depleted for the pet trade.

National and international laws now make it illegal to make pets of our own wild birds, or to import most species from other countries.

People **traffic** stolen birds under horrible conditions. 60 percent of those birds die en route, and the survivors are often placed in miserable breeding facilities similar to **puppy mills**.

If you want a pet bird, learn about the species you've chosen. A bird is a lot of work, and the bigger it is, the more complicated its care. Parrots live for decades. Buy your pet from a reputable breeder, or better yet, get it from a bird-adoption shelter. Lots of abandoned birds need good homes.

You're nothing but a jail bird!

The author warned us he has a mean streak.

Are you a Birder?

A birder is someone who loves watching birds and learning about them. Who can do that? You can!

Where's the best place to start? Your own yard, even in a city, even if you don't have a yard—you know how to find birds.

What do you need? You need your eyes, your ears, and your patience. You'll learn to check out the birds' appearance and vocalizations, their habitat and behavior—stuff we've talked about.

Start your *life list*—all the bird species you see in your entire life. Write down the bird's name, the date and place you first saw it, and other encounters and details that interest you over the years.

Get yourself a *sketchpad* and *colored pencils*.

Of course you can draw! When you draw a bird you're watching, you see colors and details you hadn't noticed before.

Save your notebooks. They'll be precious to you one day.

Can you identify the one with the purple claws?

I forgot my *Humans of the Midwest* guide at the nest.

Field guides help you identify the birds you see. In these books you'll find the birds' pictures, their names, something about them, and a tiny map that shows where they're found, winter and summer. Knowing your location on the map, you'll know which birds can be found around you. There are field guides for kids. Cell phone apps have it all; field guide, videos, recorded birdsong and they are even interactive!

Eventually you'll want *binoculars*. They magnify what you're looking at, which is great for small, faraway birds. With binoculars, a silhouetted bird on a fence becomes a kestrel. Two birds squawking over territory become Steller's jays. It's fascinating!

Kids' binoculars are great because they're lightweight and cheap, but their quality isn't always the best. Save your money for a pair that is light, easy to use, and gives a sharp image. Stores that specialize in food and products just for birds may sell decent binoculars for kids, and their salespeople can help you. Finding birds in your binoculars takes practice. Don't give up! Consult the experts. Birders love to help other birders.

Feeders, water, nest boxes, and bird-friendly plants in your yard attract birds, but it takes work to keep them up, and can be expensive.

Birding and nature organizations have programs for kids. What a great way to meet other birders, find out about field trips, and get involved in local environmental programs. And go online. You all are so computer savvy, you'll discover so much more than I can!

But you can keep it very simple. Grab your eyes and ears and a pair of binocs ("bins") if you have them, and go exploring!

Who can be a birder? You can, for your whole life. You'll never, ever run out of birds, friends to go birding with, and places to discover.

Bird food

Here are the answers to the "Who did it?" lineup on page 17. Thank the illustrator for the fun!

But first, notice that I've capitalized the birds' names in the answers below. Most birders do so when they use the bird's whole name (American Crow), but not if using its casual name (crow). I thought that was too confusing for this book, so I didn't capitalize any names. Here you can see how it's done.

A drumroll from a tree—#2, a Yellow-bellied Sapsucker.

Half-eaten hot dog—#4, American Crow, an omnivore like you.

Frog leg minus the frog—#11, the Great Blue Heron, a wader. Also #1, a falcon, the American Kestrel, who'll eat a frog if it can nab it. Its beak and claws show you it's a raptor, even though it's robin-sized. Compare its size to the eagle's.

Scratched-up garden—#9, a Galliform; this one is a Northern Bobwhite.

Pond missing its weeds and fish—#7, Canada Goose grazing the pondweeds with its serrated beak. The Belted Kingfisher #3, and Great Blue Heron #11, are snagging the fish.

No mosquitoes—#10, one of the nighthawks. They're not hawks; they hunt insects from the air. And also #6, a Northern Parula, a tiny, slender-billed warbler that gleans bugs from leaves and branches.

Sunflower seeds gone—#5, a sturdy-billed seedeater, a female Painted Bunting. (Check out the gorgeous male on page 42.)

Nectar lapped from honeysuckle—#8, a Black-chinned Hummingbird.

Here are some questions to keep you entertained.

Answers to questions 1–9 are found in the book. Answer question number 10 on your own.

1) Why are female birds often a dull color, and the males brighter?

2) Why do we hear birdsong start or increase in the spring?

3) What do flamingos and penguins have in common? What about pelicans and penguins?

4) What's the most common cause of modern-day bird extinctions?

5) Why are human siblings luckier than eagle siblings?

6) What do woodpeckers do in the spring instead of singing?

7) How is your diet like a crow's?

8) Why can't you fly?

9) What's a life list? What's a field guide? What do binoculars do for you?

10) Where can YOU go birding?

Glossary

A glossary is a short alphabetical dictionary
of words and their meanings.

A

Adapt, adaptation — Changes in a plant or animal species that give it a better chance of surviving. A change can be genetic, like webbing of the toes to help a species as swimmers, or a behavior change, like crows learning to eat McDonald's french fries. To adapt is to adjust to change.

Aerodynamic — A smooth streamlined shape that makes it easier for a bird to move through air.

B

Banding — Informational bands attached to birds' legs to help scientists understand more about that bird's life—like migration, age, behavior etc. (The bird is recaptured and the band number uploaded to a special database website to benefit science.)

Beak (bill) — Birds' lightweight toothless mouths. Beaks have many functions other than finding and eating food, such as carrying food, especially to store it or to feed babies; singing; fighting; nest-building; feather care. Beaks are different sizes and shapes, mostly depending on the bird's diet and how it finds food.

Behavior — What an animal does, like a bird flying or eating or singing.

Birds of prey — Meat-eating birds with strong talons and hooked beaks. Also called raptors.

Blueprint — A plan containing all the information necessary to make something.

Breed — When a male and female mate to create offspring. Another meaning is a group of animals of the same type, like German shepherds, which are a breed of dogs.

Buzzard — A casual term for hawks (in England) and vultures (in North America). Maybe early settlers confused our vultures with their hawks.

C

Camouflage — Feather or egg colors and patterns that blend in with the surroundings and make birds (or eggs) hard to see. Some species use camouflage to hide from their enemies (female ducks and cardinals). Other species use camouflage as they pursue prey (hawks, eagles, and owls).

Carson, Rachel — (1907-1964) A scientist and science writer. She worried about the effects of pesticides on our earth. In 1962, her book *Silent Spring* shocked many people into realizing how we were poisoning plants, animals, and ourselves. It led to the ban of the pesticide DDT in this country.

Claws   Birds' toenails. They have many functions, such as grabbing prey, climbing up tree bark, perching, preening, and fighting. Claws are different depending on what the species uses them for. Talons are especially strong claws for birds whose prey is awkward or struggles.

Collective noun   Word or words that describe a group of things, especially animals, like a *flock* of birds, a *murder* of crows, a *flamboyance* of flamingos, a *murmuration* of starlings.

Colony   Birds of one or more kinds (species) that nest or roost together.

Communal   Something done together or shared with other members of a group, like eating and roosting.

Compass   An instrument with a magnetized marker that points north and helps with direction-finding. Birds of some species have a *tiny* bit of magnetic iron in their brains that acts as a compass, helping them migrate. Amazing! By the way, a GPS (page 29) is a modern compass that uses satellites to tell you where you are. Birds don't have them.

Conservation   Saving and protecting something of value—like plants and animals, like the earth.

Cooperative   Working together for something that benefits the whole group. Birds do it.

Corvids   The noisy, feisty members of the crow family—in North America, crows, ravens, jays, magpies, and Clark's nutcrackers.

Court   To attract a mate. Behavior between male and female birds when they are in the process of forming a pair. Like dating.

Crest   Longer feathers on the top of a bird's head. A topknot.

Crop   An expandable area in the esophagus (the tube between the mouth and the stomach) for storing food. Not all birds have crops; they're common in seedeaters.

Crustacean   Shellfish like shrimp and crabs.

D

Dawn chorus   Birdsong at dawn as birds compete for territory and mates. Many birds' dawn songs are different than the ones they sing later in the day.

Dialect   A difference in the call or song of birds of the same species that live in different places. Like a "Southern" accent vs. a "New York" accent.

Display flights	Birds' fancy flight patterns that are one of many ways in which a bird communicates with another bird, especially when defending its territory or courting. Remember hummingbirds and eagles?
Diurnal	Active during the day (as opposed to nocturnal).
Droppings	Birds' poop. The dark part is like our feces (poop), and the white part is like our urine (pee), but they come out together from the cloaca.
Dummy nest	A male wren's partially built nest(s) to attract a female, advertise its territory, or perhaps to be used if the first nest is destroyed.

E

Eat crow	Having to admit an embarrassing mistake. In this case though (see "Quail"), it's a joke—a play on words.
Edge	The boundary between two different habitats, like forest and meadow. Only it's more than a boundary, because conditions may be different from either habitat. A forest edge has more sun than the middle of the forest. A meadow edge has forest cover nearby. Many birds live or hunt food along edges.
Embryo	The developing chick inside the egg.
Endanger	To put something, like people or birds or the earth, in danger. The U.S. Endangered Species Act was made a law by our government in 1972 to protect plants and animals that were in danger of going extinct.
Evolution, evolutionary	The process of genetic change in a species (not an individual bird), usually over a long period of time (millions of years).

F

Feral	Pets and farm animals and their offspring that have become wild and live on their own. Feral cats and pigs, for example.
Fertilized egg	An egg with the combined genes of both parents inside it. The beginning of a new bird.
Fledgling	A young bird who has grown its first set of flight feathers and left its nest (fledged). It is likely not able to fly quite yet.
Flock	A collective noun for a group of birds hanging out together. It may be all one species or a "mixed flock." Flock can also be a verb, to describe birds gathering together.

G

Galliforms	Turkeys, chickens, quail, and other ground feeders who flock and tend to escape predators by running as well as flying. Also called gamebirds. Some (chickens, turkeys) are domesticated (tamed).

Genes	Tiny bits of something called DNA that pass from parents to chick when the egg is fertilized. It determines everything the chick will be—its species, voice, color, etc. All living things have genes. You too!
Gravity	A natural force that brings things toward the earth's center. The heavier something is the more it is pulled toward the ground. Birds, light as they are, must overcome gravity to fly.
Grub	The larval form of an insect like a beetle. Insect-eating birds love 'em. We call our food grub sometimes.

H

Habitat	A plant's or animal's home. For example, a cardinal's habitat includes shrubs and trees. A bittern's includes wetlands, tall grasses, and water. A bird's habitat includes its survival needs, such as a safe territory, food and water, roosting and nesting sites.

I

Inherit	To receive traits that parents pass on through their genes, like feather color or hunting skills.
Instinct	Inherited behavior that a bird or animal is born knowing and doesn't have to learn.
Invertebrates	Animals without backbones, like worms and insects.
Iridescence	The play of light shining on something (like feathers or soap bubbles) and changing its colors.

L

Larva	The young of some animals, especially insects (but frogs too), that don't look or behave at all like their parents. For example, a caterpillar that becomes a butterfly. Larvae are bird food!
Lead shot	Many small lead balls that are shot from a gun instead of a single bullet. They remain in the surrounding countryside and in dead animals not picked up by hunters. When eaten by birds and other animals, they cause lead poisoning.

M

Mate	A partner. Also, a male and female joining briefly to make babies.
Migrate	To move from one place to another at the same times each year. North American birds usually migrate south in the fall to a warmer place where there is more food. They return north again in the spring to nest. **Migration fallouts** are flocks of migrating birds that descend together to ground or water after flying over large bodies of water, flying all night, or being caught in storms. These migrants are in desperate need of food and rest.

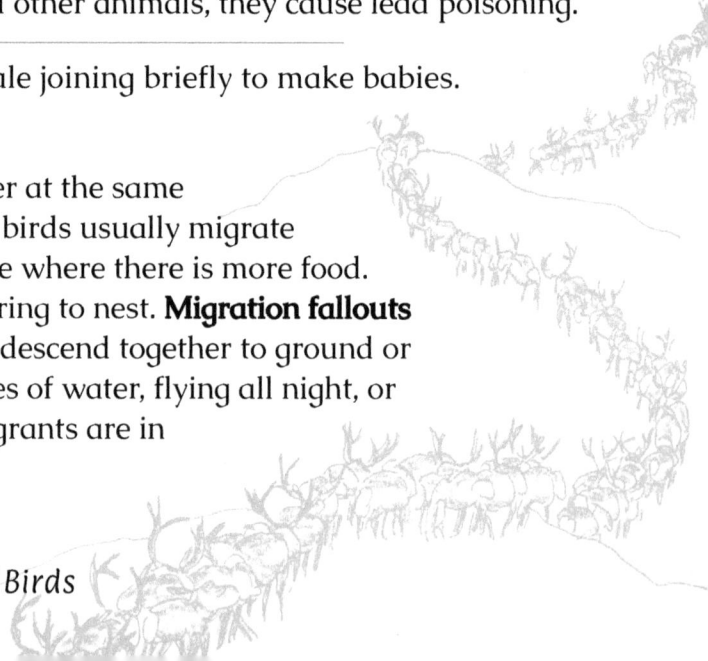

The ABZs of Birds

N

Native A plant or animal that naturally occurs in a place. Not arriving from someplace else.

Nectar A sweet liquid, made by plants and found in their flowers, that attracts insects and hummingbirds. As hummers drink the nectar, pollen falls on their heads and is spread to the next plant they visit, making it possible for that plant to reproduce.

Nocturnal Active at night. Birds like owls are awake and feeding at night, as opposed to diurnal birds that are active during the day.

O

Offspring Children.

Omnivore An animal (or person) that eats both plants and animals.

Opportunist An animal that takes advantage of a situation for its own benefit. When people are opportunists, they are often thinking only of themselves. For birds, it's a matter of survival.

P

Pelagic Birds that live near and on the ocean. Seabirds.

Pellet Parts of what a bird eats that it cannot digest. The bird regurgitates (vomits) up a nifty little package of bones, feathers, fur, beaks, claws, and insect or seed shells. Look inside an owl's pellet, and you learn a lot about its diet.

Pesticide A chemical used to kill pests, such as insects harming a farmer's crop. Unfortunately, pesticides may kill or harm many other living things, too.

Pollen Tiny powdery grains made by plants. Pollen is the "male" half (like birds' sperm) that combines with a plant's "female" half (an ovum, like a bird's unfertilized egg yolk) to make a seed, which will grow into a new plant. When hummingbirds or insects like bees and butterflies transfer pollen from one plant to another as they feed on the nectar, it's called pollination. Many, many plants rely on these pollinators to reproduce themselves.

Predators Animals that attack and eat other animals (their prey). Some birds are predators to smaller animals and prey to bigger ones. The tiny flammulated owl eats insects, and the great horned owl eats the flammulated owl.

Preen Feather care. Uses the oil from the preen gland and draws feathers, one by one, through the bill. Preening (grooming) cleans and straightens the feathers, and oils them.

Prey	An animal that is hunted and killed by another animal (a predator) for food.
Puppy mill	A business that breeds and raises puppies for sale in ways that are designed to make the most money, but are bad for the health and happiness of the pups. Happens with pet birds, too.

R

Rachel Carson	(see Carson, Rachel)
Raptors	Birds of prey with talons and strong, hooked beaks: eagles, hawks and falcons, and owls.
Roost	A place where a bird sleeps at night. It's quite common for birds of a species to sleep in the same area together — a **communal** roost. Hundreds of crows or millions of starlings may roost together.

S

Scavenger	An animal that eats dead plants and animals— and trash. To scavenge is to search for those foods.
Seed	The tiny beginning of a future plant; the equivalent of a bird's egg. It's protected inside a tough shell until it's ready to sprout (grow). Many, many birds' diets include seeds.
Sensors	A part of the body that detects something—by touching, hearing, seeing, smelling, or tasting. When birds have sensors at the end of their beaks, their beaks *feel*, like your fingers do.
Serrated	The jagged edge on the beaks of some waterfowl that helps them grasp plants or wiggly prey.
Shorebirds	Smallish migratory birds that live along our coasts or inland waters and wetlands. They have long legs for their size and are expert runners and flyers. They eat insects, small invertebrates and fish, and seeds.
Sibling	Brother or sister.
Species	A group of similar plants or animals that can breed together to make offspring. There are about 11,000 thousand species of birds worldwide. All dogs belong to one species. All people belong to one species.
State bird	All states and US territories have state birds, flowers, even fish and insects. What's your state bird?

The ABZs of Birds

Streamlined   A bird's (or plane's) shape, which makes moving through air or water easier.

Symbiosis   Plants or animals that benefit each other.

Synchronize   To arrange events so they happen together.

T

Talons   Special name given to the large, extra-strong claws (toenails) of eagles and other meat-eating birds whose prey often struggle.

Territory   An area considered "theirs" by an animal (or person or country) that is shared with friends or family and defended against trespassers. The boundaries of bird territories are invisible to humans—but not to birds.

Thermals   Air currents that occur when the ground is warmer than the air; the air closest to it warms, rises, and expands. It's a source of lift; birds take advantage of thermals to soar high without using much flapping energy.

Topknot   A cluster of feathers jutting from the top of the bird's head. A crest, a plume. Cardinals, some jays, and quail have them.

Toxin   A poison. Toxic means poisonous. Pesticides are often toxins.

Traffic   In relationship to birds, to illegally capture and sell wild birds.

W

Wading birds   Birds that wade as they search for food. They don't swim or dive. Found on shores, in marshes and wetlands. Many have long legs and bills.

Waterfowl   Collective noun for ducks, geese, swans.

Weasel   A small, skinny, short-legged, meat-eating, bird-eating mammal. As used in the Crow chapter, it's a pun, because it can also mean a cunning, untruthful person (or in this case, Crow).

Y

Yolk   The liquidy golden part of an egg; most of the egg yolk is food for the embryo before it hatches. But, almost invisible on the top of the yolk is a tiny pale spot, the *germinal spot,* that contains the genes the female is going to contribute to the baby bird. The genes in the male bird's sperm enter the germinal spot and combine with the female's genes, and right there, on the top of the egg yolk, is the beginning of new life, part mom and part dad!

The ABZs of Birds

Index

An index tells you where you can find what you're looking for in your book. The words are always alphabetized, and show the page where they're found. The page numbers are **bolded** when there's special information or the words are in the glossary, which starts on page **73**.

What do you think flamboyance means?

Some fancy thing about collective nouns

Endanger **65, 75**

Esophagus **54**

Evolution **22, 75**

Extinct, extinction 39, **64–65,** 72, 75

Eyes, eyeballs 7

F

Falcon, peregrine **57,** 78

Fall (autumn) 25, 32, 53, 67

Feathers 10, **13,** 18, 22, 23, **42–43,** 49, 62, 76, 79

Feather oil **49**

Feeders (bird) 32, 33, 71

Feeding flock **32**

Feral **46, 75**

Fertilize **63, 75**

Field guide **71, 72**

Finch, native 33

      zebra **68–69**

Flamboyances **19**

Flamingo **18–19,** 72

Fledge, fledgling **11, 75**

Flicker **60,** 61

Flight, fly 13, **22–23,** 43, **52–53** 72, 76

Flock 26, 32, **75**

Flyway **57**

Forage **5**

Fresh water 13

Friction (drag) **22**

G

Galliforms **16,** 72, **75**

Genes **63,** 75, **76**

Gizzard **54**

Gobble 54

Godwit, Bar-tailed **53**

Golf ball 21

Goose (geese) 79

      Barnacle goose **52**

      Bar-headed goose **53**

      Canada goose **38**

      Hawaiian goose **38–39**

      Snow goose 53

Gorge **53**

Gorget **25**

GPS 29, **74**

Gravity **22, 76**

Groom, grooming 17, 49

Grubs **20, 76**

Grit **54,** 55

Gular pouch **44**

Gull **20–21**

H

Habitat **46,** 56, 57, 64, 75, **76**

Hatch, hatchling **10–11**

Hawk 29, 57, 73, 78

      Red-tailed hawk **57**

Helicopter (bird) 24

Hibernate **52**

Hot dog 17, 67

Hummingbird 16, 23, **24–25** 51, 53

Hunters, hunting 26, 46, 47

I

Incubate **10,** 62

Incubation patch **10**

Inherit, inherited **25,** 37, **76**

Insect eaters 16, 53, 76

Instinct **25,** 53, **76**

Introduced birds **26–27**

Invasive **26**

Iridescent feathers **25, 76**

J

Jays **28–29**

Jelly beans 25

Junco **33**

Juveniles **53**

K

Kestrel 71, 72

Killdeer 11, **30–31**

L

Landmarks 53

Larva, larvae **13, 76**

Lead shot **59, 76**

Life list **70,** 72

Lift **22, 79**

Little brown birds **32–33**

Loon 13

Lungs **23**

She looking up weasel or looking up at us?

The ABZs of Birds

Kathryn Mikesell Hornbein, a retired pediatrician, enjoys birding, hiking, reading, and making music. She gets her kid fix by writing for children and volunteering in her local elementary school in Estes Park, Colorado.

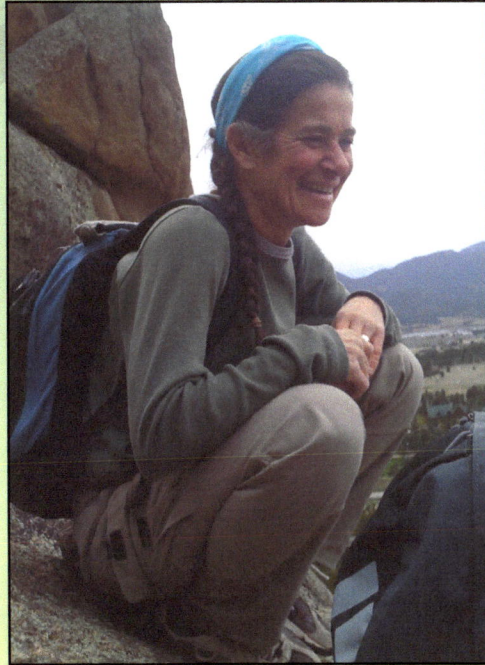

Cynthia Hunt lived in Ladakh, India in the Western Himalayas for over 30 years, where she founded nonprofits that empower women and youth in remote villages. She's written and illustrated many children's books for her development programs.

www.ingramcontent.com/pod-product-compliance
Lightning Source LLC
Chambersburg PA
CBHW050909210326
41597CB00002B/73